A SAVE–OUR–PLANET BOOK
THE PROFITS GO TO CONSERVATION

SEA ANEMONES
. . . as a hobby

TT-027

A SAVE–OUR–PLANET BOOK
THE PROFITS GO TO CONSERVATION

SEA ANEMONES
...as a hobby

U. ERICH FRIESE

© Copyright 1993
by T.F.H. Publications, Inc.

Distributed in the UNITED STATES to the Pet Trade by T.F.H. Publications, Inc., One T.F.H. Plaza, Neptune City, NJ 07753; distributed in the UNITED STATES to the Bookstore and Library Trade by National Book Network, Inc. 4720 Boston Way, Lanham MD 20706; in CANADA to the Pet Trade by H & L Pet Supplies Inc., 27 Kingston Crescent, Kitchener, Ontario N2B 2T6; Rolf C. Hagen Ltd., 3225 Sartelon Street, Montreal 382 Quebec; in CANADA to the Book Trade by Macmillan of Canada (A Division of Canada Publishing Corporation), 164 Commander Boulevard, Agincourt, Ontario M1S 3C7; in ENGLAND by T.F.H. Publications, PO Box 15, Waterlooville PO7 6BQ; in AUSTRALIA AND THE SOUTH PACIFIC by T.F.H. (Australia), Pty. Ltd., Box 149, Brookvale 2100 N.S.W., Australia; in NEW ZEALAND by Brooklands Aquarium Ltd., 5 McGiven Drive, New Plymouth, RD1 New Zealand; in the PHILIPPINES by Bio-Research, 5 Lippay Street, San Lorenzo Village, Makati, Rizal; in SOUTH AFRICA by Multipet Pty. Ltd., P.O. Box 35347, Northway, 4065, South Africa. Published by T.F.H. Publications, Inc. Manufactured in the United States of America by T.F.H. Publications, Inc.

Contents

Introduction 7
Sea Anemones as Aquarium Animals 13
Sea Anemones—A Historical Review 21
Sea Anemones as Cnidarian Animals 29
General Characteristics of Sea Anemones 33
The Biology of Sea Anemones 55
Classification of the Cnidaria 85
Subclass Hexacorallia 99
Order Corallimorpharia 107
Order Actinaria 121
Mesomyarian and Acontian Anemones 173
Subclass Octocorallia 199
Symbiosis 209
Sea Anemones and Clownfishes 231
Sea Anemones in the Minature Reef Aquarium 255
Anemone Care 303
How to Get a Sea Anemone 315
Index 317

Introduction

Flower or animal? This is commonly the question of beach-goers when they first see anemones hanging limp from harbor or pier pilings at low tide or expanded in full glory just below the low-water line. Beachcombers may find sea anemones in a rocky tide-pool, nestled in narrow cracks and crevices where small amounts of residual water protect them until the next incoming tide floods their entire habitat again.

Facing page: Many animals with tentacles can be found in the average marine aquarium, but only a few (such as the giant *Heteractis magnifica* at the top center of the photo) are true anemones. Photo: B. Kahl.

The modern marine aquarium features many cnidarians, such as hard and soft corals, as well as anemones such as *Heteractis crispa*. Photo: B. Kahl.

Divers on coral reefs will have seen giant sea anemones. If they are lucky and in the right location, there even may have been a small school of colorful clownfish living harmoniously among the filamentous, flower-like tentacles of the anemones.

The appearance of sea anemones is truly deceiving. With their graceful, petal-like tentacles swaying gently in imperceptible water currents and their elegant, often brightly colored "stems," sea anemones indeed look very much like flowers. Nevertheless, they are true members of the animal kingdom, but the primitive anatomy of these animals places them in a relatively low position on the scale of evolutionary development.

While the shapes and colors of sea anemones may be deceiving to the layman's eye, aquarists, especially those keeping marine animals, have been familiar with them for many years. Sea anemones belong to the phylum Coelenterata (sometimes also referred to as Cnidaria), a group of invertebrate animals (i.e., animals without a backbone) characterized by a simple body structure. Basically, there is a "hollow body" with only a single body opening through which food is taken in as well as the digested and undigested waste material eliminated. Closely related animals are the jellyfishes, corals, and hydras. In fact, in a somewhat over-simplified manner one could compare a sea anemone to an anchored, upside-down jellyfish.

There are about 1000 species of sea anemones, as well as a number of related animals that seem to be, in their morphology and development, positioned between true sea anemones and corals. Some of these anemone relatives will be discussed here since some of them (e.g., sea pens, tube anemones) also are popular with marine aquarists.

Sea anemones range in size from about 1 cm to 1.5 m (0.4 inch to 5 feet) in

Anemones range in size from under half an inch (1 cm) to almost 2 yards in diameter (1.5 m). The largest species are the reef-dwellers often associated with clownfishes, such as this *Heteractis crispa* with its *Amphiprion chrysopterus* guests. Photo: Dr. G. R. Allen.

In the Atlantic, which lacks truly giant reef-dwelling anemones, one of the largest anemones is the pink-tipped *Condylactis* of the Caribbean. Photo: F. Rosenzweig.

diameter. They are found in all oceans, from the intertidal zone down to a depth of about 10,000 m. Thus it stands to reason that such a vast and highly variable "habitat" must require considerable morphological, behavioral, and developmental adaptation on the part of sea anemones if they are to survive and reproduce successfully.

Some sea anemones are capable of swimming, which is an extraordinary feat for animals that under normal conditions are firmly attached to some solid object. Others "hitch rides" by living on the shells of hermit crabs and certain molluscs. Many sea anemones live on open, exposed surfaces (rocks, rocky reefs, harbor pilings, and other submarine structures), while others are typically burrowing forms that are able to withdraw rapidly into the sand or mud when danger approaches.

Such diverse habitats require that certain species be able to withstand the pounding waves of ocean surf, while others prefer the quiet backwaters of sheltered bays and harbors. Sea anemones either feed on tiny floating particles or lead a strictly predatory life, trapping small fishes and invertebrates such as crabs, shrimp, and worms among their tentacles equipped with stinging cells. All sea anemones have stinging cells within their tentacles—millions of tiny barb-like projectiles capable of being ejected with explosive force at a touch into the surface skin of prey or a potential enemy, injecting a potent venom and paralyzing or even killing the unwary intruder.

The tentacles of anemones bear numerous stinging cells, but several fishes have developed behavioral and anatomical methods of avoiding or getting used to stings. These *Amphiprion ocellaris* are at home in their *Heteractis magnifica* anemone. Photo: E. Chao.

The Caribbean *Stichodactyla helianthus* is the Atlantic representative of the Indo-Pacific giant anemones. It often has commensal shrimps of the genus *Periclimenes* present in its tentacles. Photo: C. Platt.

Sea anemones are also known for a wide range of fascinating and well-defined symbiotic relationships with certain fishes as well as with other invertebrate animals (crabs and shrimps). This topic also will be investigated in detail in this book.

Sea Anemones as Aquarium Animals

Sea anemones have been kept in aquaria for many years, apparently at least since the early 19th century. By the middle to late 1800's, there already were major public aquaria in Berlin, Hamburg, Paris, London, and other European cities that all had extensive marine life displays predominantly from neighboring seas (North Sea, Baltic Sea, Arctic Ocean).

While home aquaria have been around since well before the turn of the century, the aquarium hobby did not become widely popular until the early 1920's. First only freshwater animals were kept, but since about the early 1930's an increasing number of aquarists have devoted their attention to keeping marine animals. Sea anemones became part of the marine aquarium hobby very early on. As worldwide travel and transport systems improved during the last 30 years, so did the availability of live marine specimens.

One of the most significant early advances was the changeover from metal transport containers (insulated fish cans) to the now commonly used plastic bag/oxygen technology. This took place on a commercial scale in the early 1950's, and it went hand-in-hand with increased air transport of aquarium specimens. Not only did this improve transport efficiency and so reduce the per-specimen cost to the aquarium hobbyist, but more importantly the shorter in-transit times also assured that the condition and quality of the specimen upon arrival were far superior than in the "old days." This not only affected marine fishes but it also caused a dramatic increase in the availability of marine invertebrates, including sea anemones, to aquarium hobbyists.

There have been even more rapid advances in the support technology for marine aquaria during the last 20 years or so, most notably in seawater systems and water quality monitoring technology. Most importantly, there has been on-going research into those environmental factors that have profound affects on the well-being of marine fishes and invertebrates. Since then a better understanding of the chemistry of sea water has led to the development of better sea salts for synthetic sea water. This

The advent of high-tech equipment and techniques has changed the aquarium hobby forever. Today it is possible to keep and even breed invertebrates that only a decade ago would not survive a month. The future looks bright for the technical marine aquarium. Photo: Cheng.

Probably more giant anemones, such as this *Heteractis magnifica*, are kept by aquarists than any other type of anemones. This is partially because they host gorgeous clownfishes such as *Amphiprion perideraion*. Photo: A. Power.

has brought the marine aquarium hobby to people in areas far removed from any sea or ocean natural sea water supplies.

In recent years there also has been a growing understanding of the dynamic processes involved in the survival of marine life in the oceans, especially in those geographical regions where most of the aquarium specimens are obtained. This has led to a better awareness and understanding of those factors (salinity, currents, temperature, light, nutrient cycling and recycling) involved in maintaining an overall balanced and natural marine environment under captive conditions.

This new, comprehensive knowledge, together with the development of suitable hardware and a substantial improvement of general aquarium management practices, has culminated in what is now known in the marine aquarium hobby as "miniature reef" technology. This new technological concept and aquarium management approach has not only benefited profoundly the keeping of marine fishes, most notably coral reef fishes, but even more importantly it made keeping the much more sensitive marine invertebrates, including sea anemones, in home aquariums also quite feasible.

This book is designed to provide the marine aquarist with all he or she needs to know about keeping sea anemones. It will familiarize you with the technology and hardware required as well as with details about the biology, natural history, and basic classification of these beautiful creatures. In addition, those genera and species frequently available from aquarium shops and dealers, as well as those readily accessible through personal collecting along sea shores, will be discussed in some detail.

A well-arranged display marine aquarium can be a beautiful fixture in any home. Although the basic setup can be expensive to purchase, maintenance is relatively easy and not very expensive. Photo: Cheng.

Sea Anemones— A Historical Review

Sea anemones usually are classified together with jellyfishes, hydras, and corals in the phylum Coelenterata. In the more recent literature there is now a preference for the term Cnidaria, in recognition of the fact that 19th century usage of the term Coelenterata applied to two superficially similar but biologically distinct groups of animals—true coelenterates and comb jellies (Ctenophora). Most

Facing page: Cnidarians or coelenterates often are colonial animals (think of the huge colonies of stony corals, for instance). One of the most highly developed colonial cnidarians is the Portuguese man-of-war, *Physalia*, a siphonophore with specialized organs (actually separate animals) that produce everything from a colorful float to deadly stinging tentacles. Photo: K. Gillett.

Facing page: The jellyfish *Pelagia noctiluca* is typical of most jellyfishes, cnidarians that have a free-swimming sexual stage and a tiny sessile (anchored) asexual stage that buds off the swimming adults. Photo: R. Larson.

older references list the name Coelenterata as the preferred one, as indeed does most of the early aquarium literature. In more recent articles we find sea anemones described as being cnidarian animals, the name adopted in this book.

Like other scientists, biologists must continually revise their opinions as new evidence comes to hand or as they reach a greater understanding of a particular animal or group of animals. With that in mind, it may be of interest to follow, in an abridged form, the classification of sea anemones and related soft-bodied marine invertebrates since man first began to look seriously at the animals in the oceans around him.

Sea anemones, jellyfishes, and related forms were well known as far back as 2500 years ago. In fact, Aristotle (384 to 322 B.C.) in Greece had named these animals Acalephae or Cnidae (from the Greek *akalephe* = nettle, *cnidos* = thread).

Obviously he was well aware of the ability of these creatures to cause a painful burning sensation on the human skin when people came into physical contact with them. Yet, because of their "flower-like" appearance he apparently considered them to be in an intermediate position between plants and animals. This concept was carried forward well into the 18th century with the term zoophyta (Greek, *zoon* = animal, *phyton* = plant). Regrettably, this term also initially included most of the soft-bodied invertebrate animals, from sponges to tunicates.

The first recognition of sea anemones, hydras, and jellyfishes as a distinct group of animals came with such famous naturalists as Linnaeus, Lamarck, and Cuvier. By that stage, basically two major groups of lower animals were then established, the Radiata and the Zoophyta, though no consistency prevailed in terms of exactly what

Right: Underview of the odd jellyfish *Cassiopea*, a form with greatly branched or frilled tentacles. Photo: D. L. Ballantine.
Facing page: *Haliclystus*, a primitive jellyfish in which adults are stalked and not free-swimming. Photo: R. Larson.

should be included in each of these groups. Lamarck's Radiata included some sea anemones, as well as all starfishes, the related sea cucumbers, and also sea urchins. Yet, the remaining coelenterates (or cnidarians), such as the corals and hydras, he placed into the Zoophyta. Cuvier, on the other hand, put most of the lower invertebrates (worms and sea anemones, as well as jellyfishes) into the Acalephae or into the Polypa (corals, sponges, and bryozoans).

This system prevailed until 1829, when Eschscholtz demonstrated distinct differences between some of these animals. Subsequently he established the Ctenophora (comb jellies), the Siphonophora (polymorphic colonial jellyfishes), and the Medusae (proper jellyfishes). The first comprehensive study of a

Cassiopea frondosa often is called an "upside-down" jellyfish because it lies on its back (the bell or disc) and has the feeding tentacles projecting up into the water current. Large beds of these jellyfish carpet the bottom in areas where currents bring in abundant plankton. Photo: R. Larson.

single coelenterate species was done on the jellyfish *Aurelia* by Sars in 1829. His work demonstrated that there is a relationship between the polyp-forming coelenterates (those that are attached to a solid surface or object) and those having a medusoid (free-swimming) stage only. These had previously been considered as separate groups.

Further studies on the Zoophyta by Thompson (1831), Ehrenberg (1833), and Johnson (1838) finally caused the removal of the Bryozoa (moss animals) to their own group. In 1847, Leuckhart established the fundamental difference between the two large groups of coelenterate-like animals and the echinoderms (starfishes, sea urchins, etc.). He observed that sea anemones, jellyfishes, and hydras (freshwater polyps) use their body cavity as an intestine, so he named these animals Coelenterata (Greek, *koilos* = cavity; *enteron* = intestine).

Today's marine aquarist is able to use special high-intensity lights, exacting synthetic salt solutions, and even wave-making machinery to duplicate small sections of reef. In such high-tech tanks even corals may reproduce. Photo courtesy Dupla.

Up to that time the sponges and comb jellies were still considered to be coelenterate-like animals, and consequently Leuckhart retained these in his Coelenterata. This terminology was in use until 1888, when Hatcheck separated these animals into three phyla: Spongiaria (sponges), Ctenophora (comb jellies), and Cnidaria (the "stinging" jellies, i.e., jellyfishes, hydras, corals, and sea anemones).

Although the term Coelenterata still is frequently used, the name Cnidaria appears to be more suitable for this phylum. Some authors prefer a further subdivision into two subphyla, Cnidaria (for the coelenterates proper) and Acnidaria (for the comb jellies).

Sea Anemones as Cnidarian Animals

The most notable external body feature of all cnidarian animals is their radial or biradial symmetry. Sea anemones fall into the latter type of symmetrical arrangement. Cnidarians are sessile (attached to some solid object or surface) or free-swimming. Most are marine forms, although the hydras are well-known from the freshwater environment. The body is constructed basically of only two layers of skin, an outer epidermis and the inner gastrodermis, with some more or less well-defined but often highly variable connective tissue (mesoglea) filling the space between them.

Most cnidarians do not have significantly developed sense organs. The presence of tentacles around the mouth, used for collecting and retaining food particles, is conspicuous. Yet it is the presence of a multitude of tiny stinging cells, the cnidoblasts or nematocysts, that has given this group of animals their name (i.e., phylum Cnidaria).

These stinging cells produce one of the most complicated secretions known in the animal kingdom, a somewhat surprising fact in view of the rather primitive evolutionary status of these animals. The role of cnidoblasts is both offensive and defensive: to attack and hold prey and to fend off potential predators. Most of the stinging cells are distributed over the entire body of the cnidarian animal. They vary greatly in shape and size, and they are divided into two major categories according to their respective structure: thin, single-walled capsules referred to as spirocysts (present in only some of the cnidarians), and nematocysts proper (present throughout the phylum Cnidaria). There are 17 different types of nematocysts, based on the morphological features of the stinging cells. Sea anemones possess both types of stinging cells.

The cnidarian development features two main structural types of animal. The sessile form is represented by the polyp, which has the basic shape of an elongated cylinder. It is attached by its aboral end (the end opposite to the mouth) to some solid object or surface.

The medusa is the free-swimming type. It has a shortened body and is expanded radially into a bell-like or saucer shape.

Both of these types may occur within the same species, but at different stages of the life cycle. The definitive form in the typical cnidarian animal is either the polyp or the medusa. From an evolutionary point of view, the polyp, as the adult animal, can be described as a perpetual larval stage, while the medusa is the completely evolved cnidarian animal.

The duration of the polyp stage determines the major groups (classes) within the phylum Cnidaria. Those animals that have both a polyp and a medusoid form during their life cycle are placed into the class Hydrozoa. Cnidarian animals where the medusoid form is the main

The adult anemone is a polyp stage that is capable of sexual reproduction. This *Heteractis magnifica* is home to an *Amphiprion ocellaris* clownfish. Photo: E. Chao.

and final stage in the life cycle belong in the class Scyphozoa. Sea anemones and corals never go beyond the polyp stage, so they represent the third major group, the class Anthozoa, within the phylum Cnidaria.

The beautiful and delicate pinkish tentacles of this anemone, probably a species of *Condylactis*, belie the potent stinging cells present in every anemone. Photo: C. Platt.

General Characteristics of Sea Anemones

Sea anemones must have received their common name originally from their flower-like appearance, particularly when the oral disc (the "mouth" of sea anemones) is expanded and displays the multitude of "fluffy" tentacles. Beyond that, the major external body features of sea anemones are also conspicuously similar in general form to a plant.

The lion's mane jellyfish, *Cyanea capillata*, has a reputation as one of the largest and most deadly animals in the sea. Human deaths have resulted from encounters with large specimens. Photo: R. Abrams.

The body of the sea anemone is divided into three major parts: the oral disc, with its centrally located mouth opening surrounded by rings of tentacles; the main body, the column or "stem"; and the base (also referred to as the aboral or pedal disc), the point of attachment to some solid object or surface.

Depending upon the species, the sea anemone body may be compressed or slender and elongated. A number of species—especially the larger temperate or tropical forms—have oral discs with predominantly off-white or cream-colored tentacles (often with reddish or brownish tips and a darker colored column). Yet there also are many sea anemones that are colored in vivid yellows, reds, blues, or greens. Shades of red and yellow (orange) are common among many of the smaller species, and many species are multicolored.

The tentacles surrounding the mouth opening may form a single ring of rather thick, stout individual tentacles or several rings of sometimes very feathery, fine, often branched, tentacles. The latter give certain anemones a fluffy, carnation-like appearance.

The tentacles of sea anemones may number from a dozen rather thick ones to more than a thousand extremely small ones. But whatever the total number of tentacles in a particular species, they always occur in multiples of six (the sea

No human deaths have resulted from encounters with giant anemones such as this *Stichodactyla*, but the potential for serious stings should not be disregarded when cleaning or rearranging the tank. Photo: J. Burleson.

This group of percula clowns, *Amphiprion ocellaris*, retreat to their anemone, a species of *Condylactis* in this case, when threatened. Photo: B. Kahl.

anemones are placed systematically into the subclass Hexacorallia). Tentacles within the same ring are usually of about the same size. The innermost ring represents the oldest tentacles on a sea anemone.

The arrangement of tentacles relates to the internal anatomy of a sea anemone; specifically it corresponds to the number of internal spaces within the body cavity. These spaces or compartments are separated from each other by a special "dividing wall," the septum. Since the number of septa (plural for septum) in each sea anemone increases with age, there is also a corresponding increase in the number of tentacles as the animal grows older.

The actual shape of individual tentacles varies greatly, although they are most often of a simple

Corallimorph anemones often have tentacles ending in distinctive balls. Many are colonial, but others are solitary. This probably is a species of *Corynactis*. Photo: C. Platt.

37

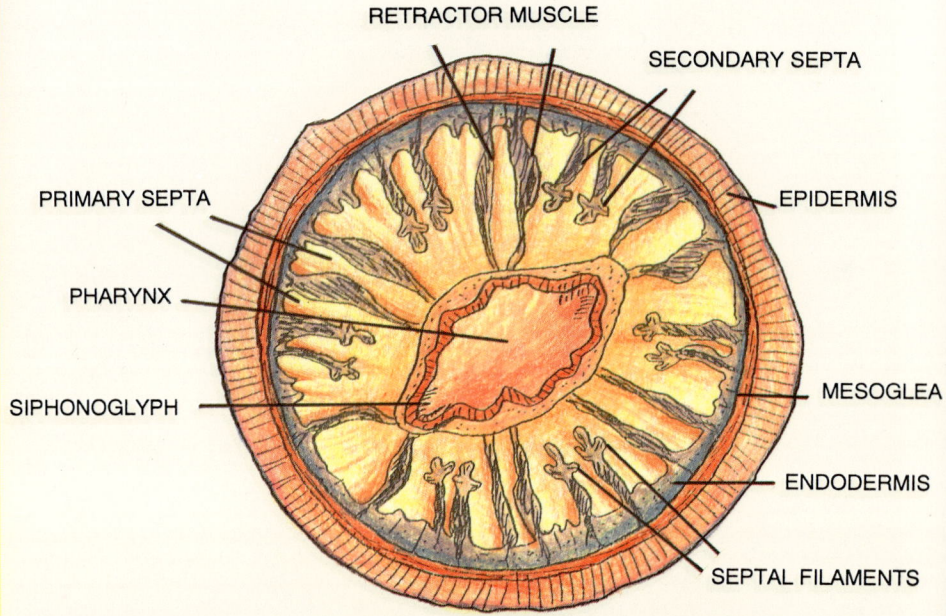

Cross section through the pharynx region of a typical anemone.

conical form that tapers to a point. In some species, partial structural restrictions in the tentacles may lead to a slightly bulbous appearance.

As mentioned above, nematocysts or "stinging cells" are the principal distinguishing feature of all cnidarian animals. In sea anemones these stinging cells cover each tentacle and often extend onto the oral disc and even onto the column. Each stinging cell consists of a single- or double-walled capsule of bulbous appearance. It contains a spirally folded, hollow thread with a minute barb at its end. The entire structure is so small that only the highest magnification will make these cells visible to the human eye.

Projecting from the outside of a nematocyst is a tiny sensor that acts like a contact point. Once this sensor is stimulated through physical or chemical contact, the stinging cell will literally explode, ejecting the tiny

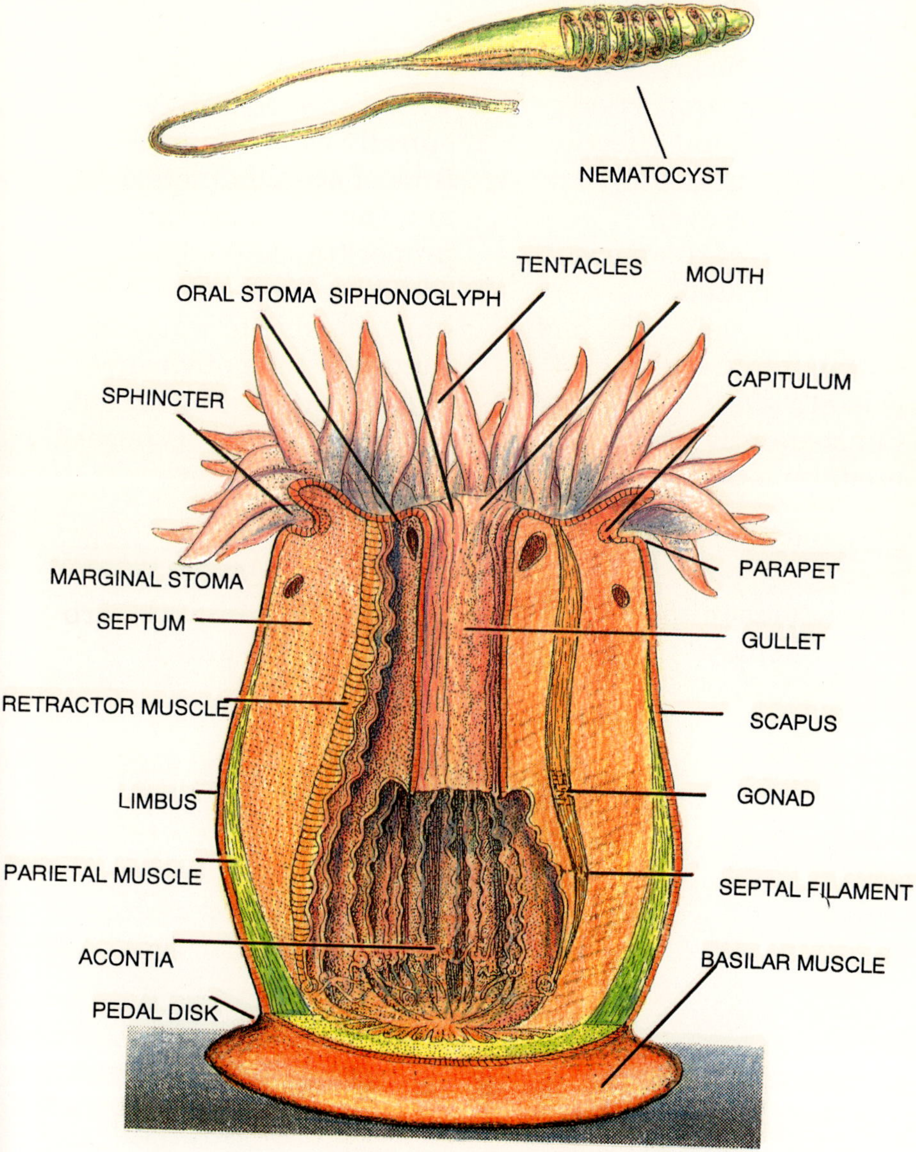

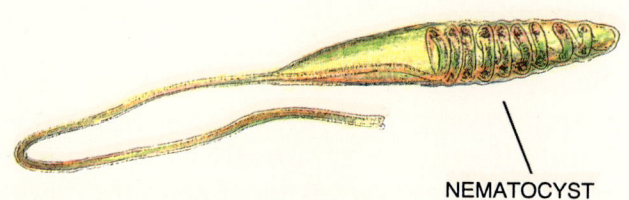

A vertical section through the column of an anemone, showing the regions of the digestive and muscular systems. A typical nematocyst is shown at the top.

barbed thread with considerable speed and force. The barb penetrates the victim's skin and simultaneously injects a potent poison. Since stinging cells are of only microscopic proportions, their individual effect is minimal, but usually hundreds or even thousands of them are being activated simultaneously, which of course multiplies the poison's effect. Consequently prey and/or predator—depending upon size—can succumb quite quickly to it.

The nematocysts of sea anemones are of two different functional types: spirocysts (single-walled capsules, permeable to water, that contain a spirally coiled unarmed tube of even diameter), and the nematocysts proper (thick, double-walled capsules that are impermeable to water and contain a tube of variable length and usually armed with spiral rows of thorns). Spirocysts often are

Anemone tentacles often are very fluid, changing shape with mood, light, and feeding condition. Some species have distinctive tentacles, such as the bulbous, white-ringed tentacles of *Entacmaea quadricolor*. Photo by Dr. F. Yasuda.

When retracted, anemones often become very odd in appearance. This seems to be a partially withdrawn *Tealia columbiana*. Photo: C. Platt.

restricted to the tentacles and oral disc. The nematocysts proper occur in wart-like batteries, some on the tentacles but most on the column of sea anemones. Sea anemone nematocysts are commonly of an elongated, slightly curved form. There are many structural types of nematocysts, but the most common one is called a basitrichous isorhiza (spirula); microbasic mastigophores and amistigophores ("bottle–brush" type or "penicillate") are also of wide occurrence in sea anemones. Hobbyists need not worry about identifying nematocyst types, of course.

These nematocysts have been studied for many years. They seem to consist largely of proteins in various chemical forms, such as tryptophane, tyrosine, arginine, alanine, glutamic acid, and aspartic

41

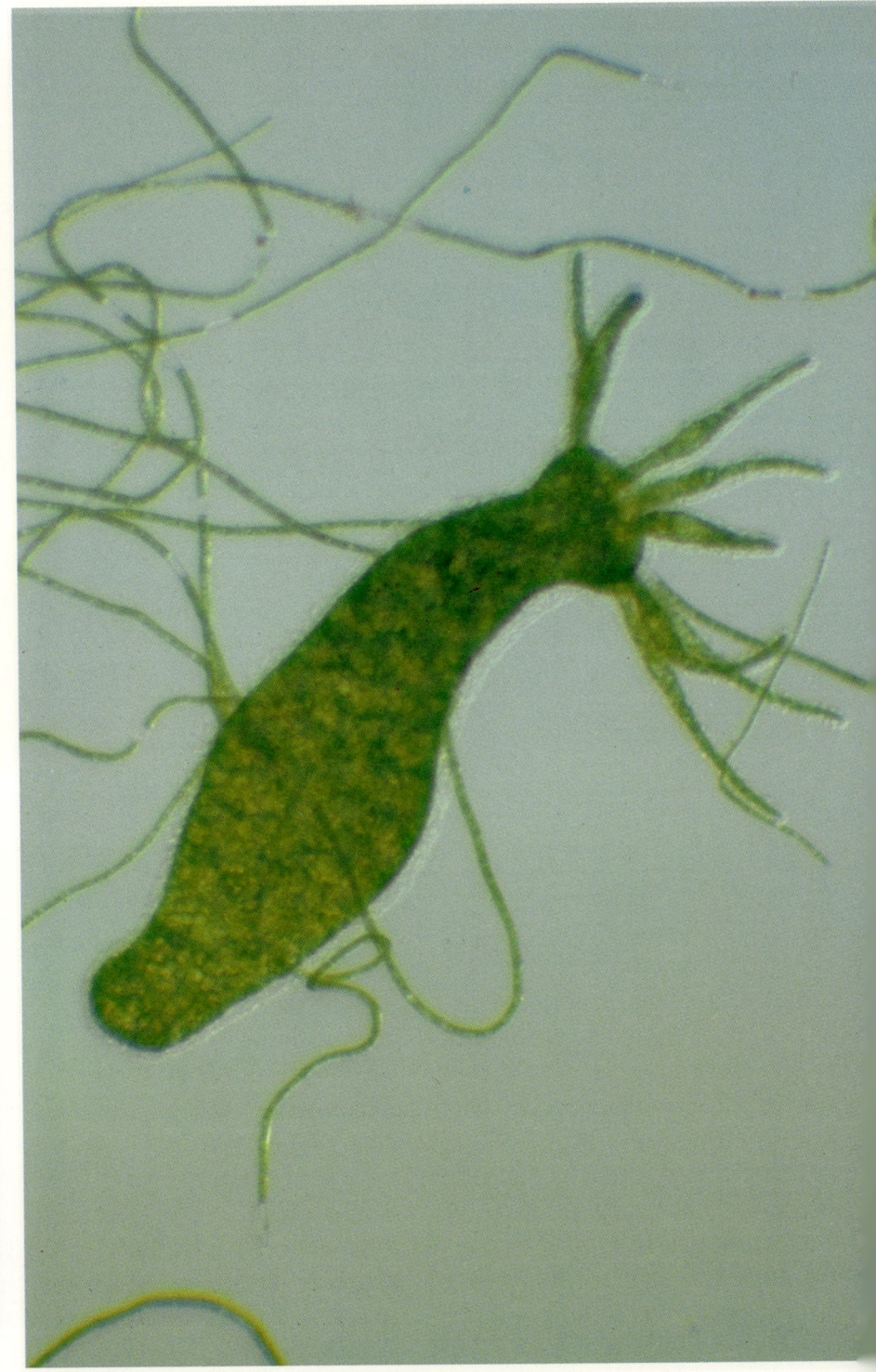

Hydras, although freshwater animals, are relatives of the anemones. Like them, their tentacles are studded with thousands of nematocysts, making hydras ferocious micropredators. Photo: D. Untergasser.

Although the nematocysts of many cnidarians can cause a rash or, in some people, allergic reactions, few cnidarians can actually kill an adult human under normal circumstances. An exception is the cubomedusa (box jellyfish) known as the sea wasp, *Chironex fleckeri*, of Australian waters. Photo: K. Gillett.

acid. Also present are sizable amounts of mineral salts, as well as several enzymes whose functions have not yet been fully determined. The actual thread appears to be of a type of keratin. It has been suggested that the energy required for discharging this thread could come from within the stinging cell due to high osmotic pressure.

The margin of a sea anemone is the junction of the oral disc and the elongated column, which in some species may be marked by a distinct groove. Depending upon the species, the column wall varies from thick and leathery to thin and transparent. Thin walls in sea anemones often permit the septa (internal partitions) inside the body cavity to show through as longitudinal lines along the column. In some species the column may be composed of both a thin wall (capitulum) at the top and a thick wall (scapus) at the lower part of the column, near its base.

The upper part of a sea anemone, either just below the margin or above the capitulum (if present), contains a pronounced thickening in the muscular layer in many species. This forms the marginal sphincter, a special muscle to close off the inverted, upper part of the anemone when it is exposed at low tide or when attacked by a predator.

The column wall may be smooth or covered with a great variety of small structural protuberances and tiny projections. Their function is not yet fully understood but is presumed to be largely adhesive and/or protective in nature. When adhesive nodules are present, the sea anemone uses them to attach an assortment of gravel, shell fragments, and sand grains to its column. This then acts as an excellent camouflage and protective cover. In a few species the column wall contains many small pores that open into tiny papillae through which water can be ejected from

the central body cavity when the animal is forced to contract suddenly.

The base or foot of sea anemones is separated from the column by a more or less distinct constriction, the limbus, from where the base expands into a circular pedal disc. With this disc sea anemones adhere to solid substrates, such as rocks, shells, or other objects in the sea. The disc also can conform to many extreme contours

This *Amphiprion perideraion* is in a host *Heteractis crispa*, one of the more distinctive giant anemones. Photo: M. Goto.

The small but spectacular "red ball anemone" actually is a corallimorph, *Pseudocorynactis*. Photo: C. Platt.

and shapes; consequently, we may find sea anemones attached by grasping a clump of mud, wrapped around stems of tube worms, or draped around the anchor chains of vessels that have been tied up for some time. In sea anemones that inhabit muddy bottoms, the pedal disc usually is just rounded or pointed, terminating in a bulbous shape to facilitate anchoring. There are a few swimming sea anemones; in these animals the entire base has become modified into a float by forming a sac-like structure filled with a chitinous network for added buoyancy.

INTERNAL ANATOMY

In order to examine the internal anatomy of sea anemones we return to the oral disc at the upper end

of the animal. There we find an oval or slit-like opening, often surrounded by a raised margin, the mouth. It is separated from the nearest tentacle ring by a smooth space, the peristome. The pharynx usually is a long and thin or short and stout tube that is slightly flattened from side to side. It conforms to the shape of the mouth. A hypothetical cut through the long axis of mouth and pharynx reveals the characteristic biradial or bilateral symmetry of sea anemones. The actual body wall is rather thick in most species, being made up of a thin outer layer of cells, the ectoderm, and a similarly thin inner layer, the endoderm. The latter borders the internal gastric cavity of the sea anemone. Between these two thin layers is a thick

The disc of an anemone or zoanthid is a study in symmetry, with the oval mouth in the center surrounded by the furrowed peristome and the tentacles. Shown is *Antiparactis*. Photo: C. Church.

The long, very slender-tipped tentacles of *Heteractis crispa* are distinctive. Often they have an "elbow" region where they bend to wave in the current. The clownfish is *Amphiprion leucokranus*. Photo: Dr. G. R. Allen.

Facing page: The fingered giant anemone, *Heteractis magnifica*, reaches a diameter of over 1 meter (about a yard) and is one of the most common reef anemones in the Indo-Pacific. It feeds well in captivity and is host to at least ten species of clownfishes, including the *Amphiprion akindynos* seen here. Photo: A. Power.

layer of featureless substance called mesoglea. It contains a collagenous fiber system similar to certain types of connective tissue found in vertebrates.

Superimposed upon the mesoglea is an endodermal ring musculature. It conforms to the shape of the gastric cavity whether it is empty or full of food or water. Some species may secrete a hard cuticle on the surface of the column and the pedal disc. Beyond that, sea anemones are devoid of any skeletal structures.

The pharynx, which can be either long or short and reduced (depending upon the species), has a series of deep, smooth-walled grooves lined by flagellated cells that beat to produce a current of water entering into the interior of the sea anemone. These grooves are called siphonoglyphs. The remaining part of the digestive system is formed by the gastric cavity proper. Its interior surface usually is rough and strongly ridged along the insertion of a series of internal membranes, attached between the internal body wall and the pharynx, extending downward into the cavity. These partitions (or septa) may reach downward along the entire wall of the cavity, or they may be only partially completed halfway down the wall. Usually both types of septa—the complete and incomplete ones—are present in each sea anemone species. In adult specimens these dividing walls always occur in multiples of six.

Just below the oral disc, close to the top of the pharynx, there is an opening in each complete septum, the oral or internal stoma. These holes permit the flow of water between the chambers formed by the septa.

Below the pharynx the complete septa curve away, making room for a large central digestive cavity (often called gastric

pouch or gastric cavity), the coelenteron. Toward the lower end of this cavity the septa again advance to the center just above the pedal disc. The free edge of each septum reaching into the cavity quite often bears a variable arrangement of small septal filaments. In some sea anemone families these filaments terminate as free threads, the acontia. They may fill up the lower part of the gastric cavity with a mass of coiled threads. Although the specific function of these acontian threads is still uncertain, they seem to play a part as a defensive mechanism of sorts. Sea anemones that are folded up and withdrawn often will have a number of these threads protruding from their mouth or through pores along their column.

A golden zoanthid, *Parazoanthus* probably *swiftii*, from the Caribbean. Zoanthids usually are small, colonial, and often found on sponges and corals. Photo: C. Platt.

The muscular system of sea anemones is well developed. The epidermal muscle is restricted to longitudinal action in the tentacles and to the radial muscles in the oral disc. A few species have a complete set of longitudinal muscles in the column and in the pharynx. The gastrodermal muscle system includes a circular layer in each of the tentacles and in the oral disc, pharynx, column wall, and pedal disc. Some sea anemones have a very special muscle, the gastrodermal sphincter, at the base of the tentacles. Its contraction will cause the sea anemone to cast off its tentacles, which appears to be a highly specialized defensive mechanism. The septa in all sea anemones are muscular; they are the primary means of contracting the entire animal.

Close-up of the expanded tips of the tentacles of the giant anemone *Entacmaea quadricolor*. The muscles in the wall of a sea anemone are well-developed and allow the animal to rapidly change shape and size. Photo: G. Spies.

If this *Amphiprion clarkii* were not immune to the nematocysts, the tentacles would easily have trapped it. Photo: U. E. Friese.

There is neither a centralized nervous system nor specific sense organs in sea anemones. Sensory perception is accomplished by special sensory nerve cells distributed mainly in the tentacles, over the oral disc in general, and in the pharynx. The base or pedal disc of sea anemones also contains substantial numbers of nerve cells.

The Biology of Sea Anemones

Sea anemones are found exclusively in the marine environment, although some of them may be able to tolerate brackish water. Many species, such as *Actinia equina, Metridium senile,* and *Anthopleura elegantissima,* can stay totally out of water for some time, until the next high tide comes. Certain genera, such as *Metridium* and *Tealia,* can live almost indefinitely in greatly

Facing page: In nature, the column of the giant anemones, such as this *Entacmaea quadricolor* with its *Amphiprion clarkii*, is hidden deep within a crevice between coral blocks, the pedal disc securely attached away from enemies. Photo: W. Deas.

diluted (brackish) sea water.

The natural habitat of sea anemones extends from the intertidal zone down to a depth of 10,000 m, possibly even deeper. Nearly all sea anemones are solitary and live attached individually to some sort of solid object or surface, but there may be many specimens in relatively close proximity. Certain sea anemones and allies *(Parazoanthus, Oulactis)* are virtually colonial, living in large aggregations and often in physical contact with each other.

While temperature places severe restrictions on the geographical distribution of particular species, this does not affect the overall range of sea anemones. They occur in all oceans and seas, from the Arctic to the Equator and to the Antarctic. The largest species are found in relatively shallow tropical water and the smallest at great depths and in very cold water. The more colorful species appear to be concentrated in the intertidal and shallow-water regions of temperate and cold seas.

Those genera with a rather wide pedal disc *(Tealia, Anemonia, Metridium)* usually are attached to large rocks, pilings, shells, and similar submerged objects. Other sea anemones prefer a distinctly muddy habitat, where their pedal disc fastens around a clump of mud or a small stone as ballast. Certain species live buried up to their tentacles in sand. Their body terminates in a conical or slightly bulbous pedal disc to facilitate adequate anchorage *(Edwardsia, Halcampa)*. In order for these species to be able to bury themselves, they bend the posterior end of the column toward the bottom. Then the circular muscle band contracts and so narrows the body or column of the animal. It then simultaneously extends its basal end into the soft substrate. Once entry has been obtained,

Burrowing tube anemones, *Cerianthus*, often live in soft, loose substrates. They protect their column and pedal area by building a parchment tube buried in the bottom. Photo: P. Wilkens.

the animal will close its mouth and force water by muscular action into its gastric pouch toward the posterior end (i.e., to the pedal disc). This will widen the initial entry hole. By repeating this process several times, the sea anemone will eventually be buried up to its tentacles in the soft bottom substrate.

Another type of burrowing sea anemone ally is the cerianthid species (tube anemones). These live inside an extended parchment tube, often buried deeply in mud, and only the spectacular tentacle crown remains visible above the substrate surface. Although they are not sea anemones in the strictest sense, they are included in this book since they are popular with marine aquarists and commonly

When a giant anemone is fully expanded on the reef, it often appears as a solid oval of tentacles. Any portion of the anemone can contract at will, however, often greatly changing the visible shape of the animal. Photo of *Heteractis magnifica* with *Amphiprion chrysopterus*: Dr. G. R. Allen.

are traded as sea anemones.

Strangely enough, there are some sea anemones (e.g., *Minyas*) that have chosen a planktonic way of life. With their oral disc (mouth) pointing downward, they float just beneath the sea surface. Their pedal disc has formed a gas chamber that contains, in addition to a gas, a spongy substance that keeps the animal afloat.

PROTECTION

Because of their inherent immobility, most sea anemones have to rely on a variety of mechanisms to avoid predation or other means of destruction, since only a few are capable of physically escaping from their enemies. Most commonly sea anemones accomplish this by withdrawing into their own protective interior. Even the slightest disturbance in the surrounding water will cause a sea anemone to expel most of the water out of its gastric cavity. Then the oral disc, including the tentacle rings, inverts into this cavity. A special musculature (the sphincter muscle) closes off the remaining opening to conserve the moisture inside the animal. Not only is the surface area of the sea anemone now greatly reduced to offer less opportunity for an attack, but this mechanism also will prevent dehydration when the anemone is exposed at low tide.

Species inhabiting sandy or muddy bottoms escape from their enemies by withdrawing completely into the protective surroundings of the substrate. The only sea anemone known to move actively away from a predator is *Stomphia*. When *S. coccinea*, for instance, is approached by the leather seastar (*Dermasterias imbricata*), it will suddenly release its foothold. Alternate lateral contractions (from side to side, about 40 per minute) of the body will propel the sea anemone away from its predator. A nudibranch,

A tube anemone, *Cerianthus*, mostly removed from its protective tube. Notice that the pedal disc is not really developed in these species, an adaptation for rapid burrowing between pebbles and sand grains. Photo: R. Larson.

The leather seastar, *Dermasterias imbricata*, is an enemy of sea anemones, as are many other seastars, but its fame comes in being a predator that actually causes the otherwise sessile *Stomphia* to escape by rapid swimming. Photo: D. Gotshall.

Aeolidia papillosa, will cause a similar reaction in *Stomphia.* Research has revealed that this profound reaction on the part of the anemone is confined to these two known predators of the sea anemone. The reaction is so strong that even a small amount of skin extract from either predator added to the water in the immediate vicinity of *Stomphia* will trigger this reaction, as if direct physical contact between them had taken place.

The stinging cells and the acontia must be viewed as active defense mechanisms. Fishes seem to be well aware of the potency of the poison discharged by the nematocysts; they can often be seen diligently avoiding contact with the swaying tentacles of sea anemones. There are, however, a few notable exceptions to this, small fishes. The column of this species has conspicuous, sac-like protuberances that are heavily armored with nematocysts. They are situated between the outer tentacle ring and the

Close-up of the madreporite of the leather seastar. In starfishes (seastars) the madreporite is a porous plate that lets water in and out of the body. The chemical scent that *Stomphia* reacts to when a leather seastar approaches probably is spread by water from the seastar's own madreporite. Photo: K. Lucas, Steinhart Aquarium.

manifested in the well-defined symbiotic relationships between some sea anemones and certain fishes and crustaceans to be discussed later.

Certain sea anemones also contain a poison within their skin. A few drops of skin homogenate of *Actinia equina* will kill column wall proper, and their shiny blue coloration is an obvious warning sign. If another specimen of *A. equina* approaches one already in place, the "resident" will inflate these sacs with water and virtually "lean" over toward the "newcomer." The skin of the sacs will adhere to the column wall

63

of the newly arrived sea anemone and will cause tissue destruction that may well lead to the death of the attacked specimen unless it retreats in time.

Other anemones with such sacs, most notably *Anemonia sulcata* and *Bunodactis,* appear to be much more tolerant toward their own kind and will permit direct contact between adjacent specimens.

In general, sea anemones serve as food for many marine animals. Among the invertebrates, certain molluscs and some

Actinia equina is an especially aggressive anemone. The column wall contains numerous nematocysts in usually conspicuous vesicles. The nematocysts will kill small fish and even other competing anemones. Photo at top: B. Kahl; photo at bottom: E. Temke.

The bright red column is typical of the Australian *Actinia tenebrosa*. This fascinating cold-water anemone bears tiny young and often lives 50 years or more in its natural habitat. Photo: U. E. Friese.

When the tentacles are withdrawn within the body, anemones often look like fleshy lumps. The European anemone above has a nearly smooth column, while the *Actinia equina* below clearly shows the shiny whitish or bluish inflated vesicles filled with nematocysts. Photos by E. Temke.

starfishes are well-known for their predation on sea anemones. Certain fishes, especially the benthic (bottom-dwelling) forms such as flatfishes, cod, and some eels, feed actively on small sea anemones.

MOVEMENT AND LOCOMOTION

In spite of their mainly sedentary and (apparently) stationary life, sea anemones may move quite frequently. In fact, some species move about a great deal. Time-lapse photography has shown that *Metridium* and *Calliactis* sway slowly and almost continuously from one side to the other in absolutely calm water without receiving any external stimuli. Such swaying motion is accomplished through alternate contractions and expansions of opposing sets of muscle bands in the column. During these movements the water pressure inside the gastric cavity may rise by as much as 7 mm (Hg). *Metridium* has the habit of expelling the entire water content out of its gastric cavity. It is then refilled through expansion of the column and by ciliary action along the oral disc. *Tealia* and other anemones move their tubular tentacles spontaneously without receiving any obvious stimulus.

Actual disturbances, such as approaching predators, may cause some species to actively move away. Some sea anemones (such as *Stomphia*) can even "swim" reasonably well. Sea anemones also may change their location because of unfavorable environmental conditions. This usually involves moves over only short distances, which are accomplished by wave-like movements of the freed pedal disc. As can be expected, sea anemones moving about in this fashion do so at extremely slow speeds. *Metridium* moves about 12 mm (0.5 inch) per hour; *Sagartia davisi* approximately 2.5 cm (1 inch) per hour. Considerably higher speeds are obtained by *Actinia equina* (almost 50 cm or 20 inches per day) and *Condylactis* (25 cm or 10 inches per hour).

Actual swimming movements have been observed in *Stomphia*. *Gonactinia prolifera* and juvenile *Stoichactis* swim by means of alternate directional beating of the tentacles. Several other species move around by alternately attaching and detaching oral and pedal discs ("somersaulting"). *Aiptasia* crawls by lying sideways along the bottom and pushing its pedal disc forward. The animal does this by contracting the circular muscle along the upper part of the body; this then pushes it forward. The gastric cavity, which is

The swimming anemone *Stomphia coccinea* as seen in Puget Sound. This attractive pastel species actually releases its grip on a rock when attacked and can swim by alternating lateral contractions of the body wall. When clear of the predator, it falls to the bottom in a suitable area and reattaches. Photo: U. E. Friese.

This colorful *Phymanthus* species from the Australian Great Barrier Reef has what are termed arborescent tentacles—small branch-like projections covered with nematocysts give the anemone additional food-catching surface area and also make it more sensitive to changes in water currents. Photo: U. E. Friese.

then inflated, anchors the pedal disc, while longitudinal muscles contract the remainder of the body and so effectively pull it up. This method permits the animal to move about 5 cm (2 inches) per hour.

Many sea anemones rely on other invertebrate animals for transport. Young specimens of *Peachia* and *Edwardsia* attach themselves to comb jellies (Ctenophora), and other anemones live on jellyfishes and certain crustaceans.

SENSORY RESPONSES

It has previously been mentioned that sea anemones have no specific sense organs, although they do possess special sensory cells that are widespread over the entire body. Large concentrations of them are found on the oral disc, especially near the mouth and along the tentacles. Consequently, it is not surprising that sea anemones respond to certain chemical, optical, and mechanical stimuli.

Some species exhibit a rather profound tidal

rhythm. When a specimen of *Actinia equina,* from the intertidal zone, is placed in an aquarium, it will follow its previous natural rhythm by contracting just before the onset of low tide in its former natural habitat. For two to three weeks a reduction in oxygen consumption can be measured in a captive specimen, coinciding with low tide in its natural habitat.

In addition to intertidal rhythms, captive sea anemones also exhibit definite day/night responses in accordance with their natural habitat. Sea anemones with a light-dependent growth of zooxanthellae (tiny intracellular algae growing inside the tissues) must

Many Mediterranean anemones, such as this snakelocks, *Anemonia sulcata*, have been well-studied both in nature and in the laboratory. Photo: B. Kahl.

The tentacles of anemones respond to chemical and thermal clues in the water and also to currents. Most can orient to face currents carrying potential prey. Photo of *Anemonia sulcata*: U. E. Friese.

have a specific amount of sunlight. Consequently, the bright green *Anthopleura xanthogrammica* and the wax anemone, *Anemonia sulcata,* will expand their elaborate tentacle crown only during daylight hours. On the other hand, typical nocturnal species, such as *Tealia felina,* will not open their tentacle rings until after dark.

Chemical response tests with sea anemones have shown a variety of fascinating results. *Metridium* can recognize dilutions of clam or mussel homogenate at a concentration of 1 ppm and will turn its oral disc toward this food source. *Actinia equina* and *Anthopleura elegantissima* actually will accept tiny pieces of paper drenched in meat juices, water soluble amino acids, and similar compounds. *Stomphia* reacts to skin extracts from its enemies.

The green color of these *Anthopleura xanthogrammica* is due to living algae called zooxanthellae. These algae need high light intensities to live. Photo: U. E. Friese.

FOODS AND FEEDING

Sea anemones are carnivorous animals requiring animal food. The size of food particles required divides them into two major categories.

First, there are the true predators, which include the genera *Peachia, Stoichactis, Calliactis, Tealia, Anemonia, Bunodactis, Actinia,* and others. These anemones have tentacles that are equipped with many nematocysts, with which they can paralyze their prey. Moreover, their tentacles are strong, suitable for grasping and manipulating food toward the mouth. They can easily overpower small fishes.

The strength of these tentacles can easily be tested by placing a hand over and against the tentacle crown. They feel "sticky," as the

Not all anemones have long tentacles. The Caribbean *Stichodactyla helianthus*, the sunflower anemone, has the disc "paved" with extremely short tentacles. Photo: U. E. Friese.

Stichodactyla helianthus can produce a mild rash in humans and often is "sticky" to the touch. Photo: Dr. H. R. Axelrod.

nematocysts are discharged into the skin of the human hand. Sensitive skin areas such as wrist and underarm should not come into contact with certain anemones, e.g., *Stichodactyla;* a serious skin rash can develop. Judging by the remains of prey found in the gut of predatory sea anemones, they usually feed on small fishes and crabs, shrimp, and other invertebrates.

The other major group of sea anemones is the particle feeders. Their food intake is facilitated by ciliary action along the outer skin. For instance, in *Halcampa* and *Sagartia* tiny cilia (actually skin cells with a long whip-like

flagellum) are distributed all over the oral disc; in other sea anemones these cilia extend over the entire body surface. Tiny floating animals (invertebrate larvae, protozoans, etc.) in the plankton will stick to these anemones by means of an adhesive produced by the ciliated cells. The cilia will then create minute directional currents toward the mouth that carry tiny food particles directly into the gastric cavity of the anemone.

One might assume that the tentacles in particle-feeding sea anemones would create a barrier to such food transport. Yet, examinations have shown that the tentacles will bend toward the mouth when their sensory cells are stimulated by food. If a small piece of paper is placed on the oral disc of *Halcampa,* it is slowly transported onto one tentacle, which then will bend away from the mouth so the paper drops off the sea anemone. If, however, this piece of paper is soaked in mussel juice, the transport system reverses

In the truly gigantic *Stichodactyla mertensii* (1.5 m in diameter), the tentacles are not "sticky" and do not adhere to human flesh as in several related species. Photo: U. E. Friese.

Facing page: An expanded specimen of *Rhodactis howesi* looks a bit like a true giant anemone, but the anemone is really a corallimorph, one of the mushroom anemones. Photo: K. Gillett.

and the paper will soon disappear into the mouth of the sea anemone.

In addition to feeding on a variety of prey, many sea anemones (and most corals and related forms) derive a considerable amount of their nutritional intake through a symbiotic association with a group of dinoflagellate algae called zooxanthellae. Exactly what is provided by the algae and in what form still are largely unknown. In any event, sea anemones with massive zooxanthellae aggregations can live for prolonged periods of time without supplementary food, provided adequate light for photosynthesis is available.

Sometimes it is not easy to determine which sea anemone is a predator and which is a particle feeder. This is a critical point for marine aquarists, since they have to know what food to offer, and an interesting case study is submitted here. Mushroom anemones of the family Actinodiscidae (order Corallimorpharia) are becoming increasingly popular with marine aquarists, especially those who maintain well-established miniature reef tanks.

Generally two genera are available through aquarium shops, the small, colonial *Actinodiscus* and the much larger *Rhodactis,* also often referred to as "elephant ears."

In spite of the fact that the Actinodiscidae have massive numbers of zooxanthellae in their tissue, both genera mentioned above have always been assumed to be primarily particle feeders. Now news has come to hand showing that a *Rhodactis* species is known to have preyed upon a number of larger fishes, including *Dascyllus,* a small boxfish, various damselfishes (but not *Amphiprion),* and some shrimp species. So, marine aquarists beware!

The food is digested inside the gastric cavity,

Facing page: Anemones often serve as host to commensal shrimp such as this female *Periclimenes brevicarpalis*. They feed on detritus from the anemone's meal and also may eat the nutrient-rich mucus secreted by the anemone. Photo: A. Norman.

and the remains are then excreted through the mouth opening to be carried off and discarded by the tentacles.

REPRODUCTION AND DEVELOPMENT

Sea anemones are either hermaphroditic (the same animal having both male and female sex glands) or are dioecious (the male and female being separate animals). The gonads occur as thickened bands along the septa in the gastric cavity. During sexual reproduction the male releases its sperm through the mouth opening, the tips of the tentacles, or through special pores along the column. In many species sperm release will stimulate females to discharge their eggs simultaneously. It has been observed that the females of certain anemones (e.g., *Halcampa*) will actually lean toward a male or even crawl closer to a male to facilitate better fertilization of the eggs.

In *Sagartia troglodytes*, male and female join their basal discs to form a space into which eggs and sperm are released. Perhaps most sea anemones (e.g., *Metridium, Anemonia*) will simply release eggs and sperm into the water, where fertilization takes place. A different mechanism exists in *Halcampa* and *Actinia,* where the eggs are fertilized inside the gastric pouch. *Halcampa* produces a thick film to surround the entire egg mass, which in turn is then released onto the substrate, where small grains of sand adhere to the egg mass and anchor it down. *Actinia* species incubate the eggs inside the gastric pouch, from which the young eventually will be released when fully developed.

There also are latitudinal variations of the reproductive behavior within the same species. In high latitudes of the North Atlantic, *Tealia felina* incubates its eggs in the gastric cavity. Further

Aquarists have to know if their anemone is carnivorous and likely to eat their fish (except adapted clownfishes such as the *Amphiprion melanopus* show here in its host *Entacmaea quadricolor*) or prefers to feed on fine particles, plankton, and the chemicals produced by zooxanthellae. Photo: Dr. G. R. Allen.

The various species of *Tealia* (shown is *T. lofotensis*) vary in how they produce their young. Most shoot eggs and sperm into the ocean, but others incubate their eggs internally or even produce tiny young in the disc. Photo: U. E. Friese.

south, the same species discharges its eggs directly into the surrounding water to be fertilized and develop in the open ocean.

Typically, the sea anemone eggs will hatch into tiny planula larvae. With the help of bands of cilia these larvae actively swim among the plankton (*Metridium, Anemonia*). The young of *Halcampa* hatch after about three weeks as larval forms without cilia and then attach themselves immediately onto solid objects. The larvae of *Peachia* adhere to jellyfish. Once the larval stage has been completed, the animal will settle to the bottom and become sessile as a small but fully developed sea anemone.

Asexual reproduction usually is accomplished by longitudinal splitting of the adult animal. A furrow develops on each side of

the column perpendicular to the plane of symmetry of the anemone. It becomes progressively deeper and eventually splits the animal. This process may take from a few hours *(Anemonia)* to a few days or even weeks. Some anemones undergo both sexual and asexual reproduction. Asexual reproduction commonly is achieved by means of laceration *(Metridium)*. A small piece of tissue is shed from the pedal disc. This develops in a relatively short time into a tiny sea anemone.

Snakelocks anemones, *Anemonia sulcata*, often reproduce by fission, longitudinal splitting of the disc and column into two complete animals. Photo: R. A. Patzner.

83

Classification of the Cnidaria

PHYLUM CNIDARIA—CNIDARIAN ANIMALS WITH STINGING CELLS. ABOUT 9,000 SPECIES IN THREE CLASSES.

I) Class Hydrozoa—Cnidarian animals with both polyp and medusoid forms in their life cycle. About 2,700 species (including 700 free-swimming medusae) in three major orders.

II) Class Scyphozoa—Cnidarian animals with the medusoid form as the main stage in their life

Facing page: Classifying anemones and other cnidarians is difficult at best because so many characters are strictly internal. Tentacles (like the distinctive organs of *Entacmaea quadricolor*) contract in preservative and lose their color patterns. Anemones must be anesthetized before preservation so the features can be seen. Many features, such as types of nematocysts, can be seen only on microscope slides. Photo: G. Spies.

Many anemones, such as this *Tealia lofotensis* from California (shown eating a seastar), have distinctive features on the column, including swollen vesicles and unique color patterns. Photo: D. Gotshall.

cycle. About 200 species in five major orders.

III) Class Anthozoa (Sea anemones, corals, sea pens)—Cnidarian animals with permanent polyp stage. About 6,000 species.

Subclass Ceriantipatharia—Tube anemones.

Order Ceriantharia—Burrowing, solitary sea anemones inside parchment-like tubes. About 50 species.

Subclass Octocorallia—Soft corals, gorgonians, sea pens.

Order Pennatularia—Sea pens.

Subclass Hexacorallia—Mushroom corals, brain corals, various solitary corals, reef-forming corals, sea anemones, encrusting anemones.

Order Zoantharia—Encrusting anemones.

Order Scleractinia—Stony corals, including mushroom corals, brain corals, and reef-forming species.

Order Corallimorpharia—Coral anemones, mushroom anemones.

Order Actinaria—True sea anemones.

Tribe Abasilaria—Sea anemones without disc muscles, the aboral end rounded, swollen.

Tribe Boloceroidaria—Sea anemones without disc muscles, the aboral end not swollen.

Tribe Endomyaria—Sea anemones with well-developed pedal disc.

Tribe Mesomyaria—Sea anemones without acontia and abdominal pores.

Tribe Acontiaria—Sea anemones with strong acontia.

Arrangement of tentacles often is distinctive for a genus or family of anemones, but seldom meaningful at the specific level. Note the alternating rows of inner tentacles in this *Tealia coriacea*. Photo: U. E Friese.

GENERAL CONSIDERATIONS

One of the most notable features of sea anemones is their highly variable coloration, not only between different species, but even within particular species. For instance, the brilliant green *Anthopleura xanthogrammica* from the Northeast Pacific can also be snow-white when occurring in a very dark habitat (because no zooxanthellae are present). Similarly, *Tealia coriacea*, common in Puget Sound, displays highly variable

Anemones depending on zooxanthellae for their coloration may vary from brilliantly colored to purest white. If the zooxanthellae are lost (due to weak lighting, for instance), the anemone's color disappears. This is *Anthopleura xanthogrammica* from California. Photo: D. Gotshall.

patterns of red, brown, yellow, and tan. It is this intraspecific color variability that makes it very difficult at times to classify sea anemones on the basis of external characteristics.

The coloration of sea anemones is based predominantly upon concentrations of carotenoid pigments that within a particular species may combine with certain proteins to produce different colors. The previously mentioned group of intracellular dinoflagellate algae (zooxanthellae) play a significant role in the colors of sea anemones. The resultant color variations within the same species, together with the fact that some families and species occur over extremely large geographical areas, have made it rather difficult to correctly identify many

Probably no two specimens of *Tealia coriacea* are colored exactly alike. This is typical for almost all anemones, which are among the most variable of animals in obvious external characters. Photo: U. E. Friese.

species—even entire genera and families of sea anemones—on the basis of coloration alone.

Over and above this, the problem of correctly identifying sea anemones and related forms is further compounded by the fact that many classes and orders have not yet been clearly delineated. Therefore, the eager marine aquarist may find different taxonomic schemes for the same animals in different books, even in those that came out at about the same time.

Apart from the lack of widespread specific knowledge, for instance about sea anemones, there also are two different approaches used by some authors for establishing taxonomic groups. First, there is a trend toward establishing a practical system. Here the taxonomic units (species, genera, families, orders, etc.) are based on character combinations, irrespective of the evolutionary development of such taxonomic units. The second approach utilizes the phylogenetic (evolutionary relationships) system, which is not necessarily reflected in the superficial (externally visible) characteristics of an animal. There are some major problems with the latter approach. Considerable difficulties frequently are encountered when it comes to identifying and categorizing certain parallel developments among widely different groups of animals (= convergent evolution). Such convergences are often extremely difficult to identify as such. Consequently, there is an ever-present danger that, because of such convergent characteristics, species that are vastly different in terms of their evolution will be placed into the same taxonomic unit.

The classificiation of sea anemones listed in this book follows the phylogenetic approach, rather than arbitrarily

lumping together the combinations of characteristics of the various species. Therefore, it appears intuitively more practical and in the light of (scant) available evidence more realistic to use tribes rather than distinct families when dealing with those sea anemones that are likely to be at least superficially similar. This makes it considerably easier to deal with the anemones in an orderly fashion.

SUBCLASS CERIANTIPATHARIA, ORDER CERIANTHARIA (Tube Anemones)

Although these animals look like true sea anemones, they differ

Tube or cerianthid anemones can be recognized by the long, slender, non-retractable tentacles, those near the mouth often very short. Photo of *Cerianthus*: Dr. D. Terver.

substantially in their body structure and early life history from the mainstream of anthozoan animals. Phylogenetically they are considered to be the original (earliest) anthozoans, so it appears to be justified to place them into a separate subclass ahead of all other anthozoans. They are popular marine aquarium animals, so they deserve to be considered here in some detail.

Cerianthid (or tube) anemones are solitary and nearly always sessile cnidarians that occur predominantly in tropical, subtropical, and temperate seas. A few species are even found in the cold water of the high northern latitudes. They are characterized by the absence of a defined pedal disc, which is not required since they are adapted for burrowing vertically into soft sand and mud. There they live in parchment-like tubes that are coated with sand grains, shell fragments, and other debris. This tube is secreted by the animal's smooth body wall. Also absent is the marginal sphincter muscle, so the long, wavy tentacles can not be withdrawn into the body cavity. Instead, the animals simply retract inside the tube. Some *Cerianthus* burrows can be up to 2 m (6 feet) long!

The extremely thin and structurally simple tentacles are short around the oral opening but very long around the margin. They occur in one to four circular rows. A *Cerianthus* about 2.5 cm (1 inch) thick may have tentacles that cover a 20-cm (8-inch) radius around the anemone. The tentacular crown of these anemones is indeed spectacular, and it is for that reason that they are very popular as miniature reef aquarium inhabitants. The colors are primarily in shades of light gray to dark brown and even greenish. Some species are nearly violet.

Cerianthus membranaceus, the Mediterranean tube anemone, is typical of its group. A solitary animal, it prefers soft bottoms in calm water where it can build its tube without problems.

Cerianthus membranaceus (Mediterranean Tube Anemone)

This is the species most commonly kept by European marine aquarists. It reaches a maximum length of about 40 cm (16 inches) and occurs in most regions of the Mediterranean Sea northward around the Iberian Peninsula into the English Channel. *C. membranaceus*, like all other species in this order, prefers calm water, where it lives burrowed into a muddy bottom, often among eel grass patches. This can be in bays, harbors, and even river estuaries. The animal is not particular about water clarity and has been found under very murky conditions.

The body is tubular and carries 144 thin, generally long tentacles, distributed in four circles from the outer margin toward the oral opening (where they are shortest). Usually *C. membranaceus* is found down to depths of about 10 to 15 m (30-45 feet), depending on the ambient water clarity. It lives inside burrows that can be as long as 1 m (40 inches). When approached by a diver the animal will

Facing page: The Atlantic tube anemone, *Cerianthus borealis*, is the eastern North American version of the Mediterranean tube anemone. In nature the tube would be more completely buried.

retract into the tube and burrow with lightning speed. This makes these animals very difficult to collect because they have to be excavated, tube and all, with a shovel or spade. Great care must be taken not to damage the animal.

Because of the long tube, *Cerianthus* requires a tank with a VERY DEEP substrate. Moreover, a sharp, angular substrate (coarse gravel), which could damage the tube, must be avoided. Instead of mud, *Cerianthus* is known to accept very fine coral sand in captivity. The color of adult specimens can vary from light gray and light brown to violet and green.

Cerianthus borealis (Atlantic Tube Anemone)

This is a slightly smaller species than the preceding one. It has only 64 marginal tentacles in four circular rows. It occurs from northernmost Arctic waters along the eastern coast of Canada, southward to Cape Hatteras.

Cerianthus lloydii (Northern European Tube Anemone)

The range of this species represents the continuation of *C. borealis* across the Arctic toward Europe. It is commonly found along the western coastline of northern Europe. All *Cerianthus* species are particle feeders, utilizing primarily small crustaceans (copepods) in the wild. In captivity they will readily accept a number of substitutes, such as frozen or fresh mussel and fish flesh, small shrimp, and frozen brine shrimp. Caution: Do not overfeed.

Ceriantheopsis americanus (North American Tube Anemone)

This species represents the southward extension of *C. borealis* along the North American continent. Its center of abundance is southern Florida and the northern Gulf of Mexico, and it extends northward as far

as about Cape Hatteras. There does not appear to be any mixing or geographical overlap with *C. borealis*. The largest tube anemone is the giant *Ceriantheomorpha brasiliensis,* from the Gulf of Mexico and the central Caribbean south.

When properly maintained in a miniature reef aquarium, cerianthids can reach considerable longevities. They are known to have lived in captivity for more than 30 years! During reproduction they produce eggs that develop into a free-swimming planktonic larva that eventually metamorphoses into the bottom-dwelling adult form. It settles down to the bottom, immediately begins to form a tube, and burrows into the muddy substrate. Reproduction through laceration also is common. Even small pieces of tissue can develop into a new animal.

An attractive orange tube anemone, possibly *Cerianthus lloydii*. Photo: G. Spies.

This view of *Ceriantheopsis americanus* clearly shows the small inner tentacles surrounding the mouth and contrasting greatly in size and shape with the several outer circles of "normal" tentacles. Photo: B. Kahl.

Pachycerianthus torreyi is one of several tube anemones found along the East Pacific coast. Identification of tube anemones, like that of typical anemones, is complex. Photo: K. Lucas, Steinhart Aquarium.

Pachycerianthus torreyi (Eastern Pacific Tube Anemone)

This species, together with *P. estuari* and *P. johnsoni,* is found along the West Coast of North America. All three are sometimes kept by marine aquarists, and they also are frequently on display in public aquaria. For details about aquarium maintenance see *Cerianthus membranaceus.*

Subclass *Hexacorallia*

These cnidarian animals usually possess six (or multiples of six) septa. The tentacles rarely are branched, and the gonads are essentially flat against the inside of the body cavity wall.

ORDER ZOANTHARIA (ENCRUSTING ANEMONES)

There are about 300 species of these small sea anemone-like cnidarians.

They occur primarily in large colonies and thus can cover, sponge-like, extensive areas. Each polyp is joined at its base to the ones adjacent to it. Their musculature is weak. The tentacles are located around the outer margin of the oral disc in two concentric rows.

They can be found in coastal waters (shallow as well as deep) in tropical as well as cold seas. Their center of abundance appears to be the tropical western Atlantic and the Caribbean. All species are small particle feeders. Most are heavily endowed with symbiotic algae (zooxanthellae), and if they are provided with sufficient light of the proper (violet/blue) wavelength (actinic 03 lighting) they need not be fed at all in a miniature reef aquarium.

Members of the family Zoanthidae often are imported for the marine aquarium trade. Most notable is *Parazoanthus axinellae* from the temperate eastern Atlantic (North Africa to the English Channel), which is golden yellow, orange, or even brownish. Also often imported is the gray and tan *Epizoanthus arenaceus* from North Africa and the northern Mediterranean. Both polyps are only about 1 cm (0.4 inch) tall and grow in vast patches along rocky shores, usually well below any wave and surf action. *E. arenaceus* often forms a symbiotic relationship with hermit crabs. Several *Epizoanthus* species commonly are sold in aquarium shops. Because of their size and location they always must be handled and kept as colonies. Individual polyps must never be removed. Instead, entire patches of these anemones should be very carefully and gently removed with a hammer and a rock chisel as a complete colony still attached to the rock. Because this type of collecting activity is destructive of the habit, it is frowned upon or even illegal in some areas. Check before you collect!

Zoanthus sociatus (Green Mat Anemone)

This is essentially a shallow-water zoanthid that occurs in water of less than 5 m (15 feet) depth. It is common throughout the Caribbean, where it forms lush green mats mostly on the inside of shallow reef flats. The oral disc is surrounded by about 30 short, relatively thick tentacles forming two rings. Adult polyps range in size from 2 to 5 cm (0.8-2 in) in height. This species, like all other zoanthids, will not require much planktonic food in captivity; however, because of the presence of symbiotic algae, adequate light is essential.

Palythoa caribbea (Golden Mat Anemone)

This species, which also is widespread throughout the Caribbean, prefers

Pieces of living rock often contain small colonies of zoanthids. Never attempt to separate the animals of a zoanthid colony in order to transplant a new colony—the polyps are connected and will die if removed. Art: J. R. Quinn.

Facing page: Zoanthids often are found on living or dead sea whips, where they may spread to cover the entire animal. Zoanthid colonies in turn offer shelter and access to food to other animals, such as the brittlestars seen here. Photo: Dr. L. P. Zann.

shallow water and moderate exposure to gentle wave action. In captivity this will have to be simulated either with a small dump bucket arrangement or—at least—with a periodic/automatic level siphon (to create artificial tidal action). Like most zoanthids, this species grows in vast colonies, forming golden brownish mats.

Palythoa grandis (Mushroom Mat Anemone)

Unlike the preceding two species, this one does not grow in large colonies. There usually are only individuals or small groups of up to a dozen specimens. It is characterized by a somewhat convoluted oral disc with very short and fine tentacles along the irregular outer margin. The color of the oral disc is usually a dirty olive-green, and the outer disc margin, including the tentacles, is light grayish brown. It has been described as having the appearance of a flattened mushroom. This is a deep-water species that occurs at depths from 20 to 40 m.

Parazoanthus swiftii (Yellow Sponge Anemone)

A few zoanthids form symbiotic relationships, mainly with sponges. This species usually is found closely associated with the branching sponge *Iotrochota birotula*. Since the zoanthid is yellowish and the sponge normally is dark green, there is a spectacular color contrast between them. This is clearly a visual warning to predators. The zoanthid contains a potent poison that discourages potential sponge predators (such as the rock beauty, *Holacanthus tricolor*). As many as 200 zoanthid polyps may cover a large sponge, forming bands around the branches of the sponge. Common throughout the Caribbean.

Parazoanthus parasiticus (Chalky Sponge Anemone)

This is another symbiotic zoanthid that commonly is found living in association with sponges. It occurs in various locations throughout the Caribbean and as far north as Bermuda. Unlike the previous species, it live in small groups evenly spaced over the sponge host. Calcareous (coral) sand becomes encrusted in the column of each polyp, which gives it a whitish coloration. Massive zooxanthellae aggregations in the tentacles color them primarily brownish.

Parazoanthus axinellae (Mediterranean Mat Anemone)

Apart from sponges, this zoanthid also lives with ascidians as well as growing over rocky surfaces. Frequently it is found on vertical and overhanging cliff walls. It occurs at depths from 1 m (3 feet) down to 240 m (750 feet).

Facing page: Zoanthid colonies may be large, with dozens of polyps (top photo) or smaller, with only a couple to a dozen or so animals. Golden zoanthids of various types often are imported. Shown at the bottom is *Parazoanthus axinellae*. Photos: U. E. Friese.

Left: The corallimorph *Ricordea florida* often is mistaken for a zoanthid. Photo: P. Colin.

Typical or true anemones, *Actinia* sp. True anemones are separate animals, not close colonies, unlike the corallimorph anemones now so commonly kept in the aquarium.

Order *Corallimorpharia* (CORAL ANEMONES, MUSHROOM ANEMONES)

The group of anemones formerly referred to as "actinarian" sea anemones in general now has been further subdivided into two groups. There now are the "true" sea anemones in the order Actinaria and "coral" anemones in the order Corallimorpharia. Corallimorphs display certain similarities to stony corals (newly formed polyps are, like corals, not separated, but there is no calcareous skeleton formed). This order contains two families, ten genera, and about 40 species.

FAMILY CORALLIMORPHIDAE (JEWEL ANEMONES)

All members of this family are small, colony-forming, and have tentacles that have club-shaped or spherical enlargements at their tips. Most species are temperate and tropical. The family is of worldwide distribution and has three genera and about 22 species.

These delightful little anemones are easily kept in marine aquaria. Much of the required nutrients appear to be provided by the zooxanthellae living within the tissues of these anemones. Provided the tank has sufficient illumination, particularly in the violet to blue spectrum (actinic 03 lighting), they will do quite well in captivity without supplementary feeding. Some species occur in fairly dark habitats (caves, under large boulders) that still are accessible to light in the 380 nm to 450 nm range of the spectrum. Most species live close to the low-water mark, but usually in the proximity of wave and surf action. While the coloration within a species can be highly variable, it is identical within each colony.

Successful aquarium maintenance of these anemones seems to depend largely on supplementary wave and tidal simulation. *Corynactis australis* has been kept for years at the Taronga Zoo Aquarium in Sydney, yet it never did very well until tidal and wave action was simulated by an automatic, constant-level siphon and dump bucket. Then vast colonies of this delightful little sea anemone developed very suddenly and were kept for many years at the Aquarium.

Because of their bright colors and relatively easy

maintenance, these anemones make excellent inhabitants for the miniature reef aquarium. In order to transplant these colonial animals successfully into a marine aquarium, a piece of rock will have to be chopped off to form the basis for a new colony. This destructive collecting is frowned upon in some areas and may be illegal—check first.

Corynactis californica, the California jewel anemone, exhibits the balls at the tentacle tips typical of its family. Photo: D. Wrobel.

Corynactis australis (Australian Jewel Anemone)

This small yellowish to orange sea anemone occurs in vast colonies around the southern half of Australia, with the exception of the southernmost tip of the island continent. Its tentacles are white with typical club-shaped tips. *C. australis* remains below the low-water mark and can occur as deep as 20 m (60 feet). Its greatest abundance seems to be in close proximity to wave and tidal action, yet it prefers protection against the direct physical impact of waves, so it commonly is found on the leeward side of large boulders and cliff faces, just below the low-tide level. Usually it does not require supplementary feeding.

Corynactis viridis (European Jewel Anemone)

This species ranges from the northern Mediterranean Sea to the northern parts of the North Sea (Scotland). It lives in vast colonies, primarily in relatively dark areas, often in conjunction with other lower colonial invertebrates. Its colors are spectacular, ranging from dark green to yellow, crimson red, and violet. This species usually does not require supplementary feeding.

Corynactis californica (California Jewel Anemone)

The predominant color of this attractive little sea anemone is shades of dark orange to bright red, with white, club-tipped tentacles. It is a colonial anemone that occurs from northern to southern California, possibly extending well along the Pacific coast of Baja California and beyond. It lives in habitats that do not get direct sunlight and often is seen fully expanded at night.

Virtually all major geographical oceanic regions have representative species of

Like many other corallimorphs, *Corynactis californica* is likely to expand fully only at night. Normally colonies are found in dark crevices and other areas not exposed to bright light. Photo: U. E. Friese.

jewel anemones. All those species available commercially in the aquarium trade seem to do well in the aquarium, provided the conditions of lighting and water activity are being met.

Ricordea florida (Knobby Jewel Anemone)

This species is widespread throughout the Caribbean north to the Bahamas, where it can be found on the leeward as well as windward sides of a reef, from just below low-tide level down to a depth of about 20 m (60 feet).

It is one of the most unusual looking sea anemones. Vast numbers can be seen densely crowded and virtually encrusting large objects or surfaces under water, and it then often is quite difficult to distinguish individual anemones. Although adult specimens can measure to about 5 cm (2 inches) across the oral disc, the column is only about 2 cm (0.8 inch) tall. An unusual feature is the massive aggregation of extremely short, rounded tentacles over the entire oral surface. The color of *R. florida* is highly variable, depending upon

A splendid marine aquarium in which plants, mostly the alga *Caulerpa*, form the backdrop for a few tiny fishes (*Amphiprion perideraion* and *Centropyge acanthops*) and a large, unidentified anemone. Photo: B. Degen.

the habitat and prevailing light level. The outer areas of the tentacles usually are different shades of green with a contrasting color at the base. The oral opening and the outer ring of tentacles may be a different color from the rest of the disc.

The aquarium literature does not appear to show any record of this species being kept in captivity. From the appearance and habitat description it can be assumed that adequate light levels and proper water conditions in a miniature reef aquarium should be sufficient to provide optimal conditions in captivity.

FAMILY ACTINODISCIDAE (MUSHROOM ANEMONES)

These sea anemones are characterized by an extremely broad and flat oral disc relative to a fairly narrow column. The tentacles are extremely short (in some species wart-like) and non-retractile. They are arranged in radial rows or radial patches. The column wall is smooth, and there is a broad, adhesive pedal disc.

In early aquarium and related literature some of these sea anemones had been confused with the superficially similar but unrelated giant tropical sea anemones of the genera *Stichodactyla* and *Heteractis*.

Mushroom anemones have been given their common name from their Chinese mushroom-like appearance. They have become very popular with marine aquarists in recent years. Mushroom anemones are quite distinctive and highly attractive since they come in a wide range of colors from reddish brown to bright green and metallic blue, and many shades in between. It is this variety in prevailing colors that makes it difficult to identify different species. So, it is not surprising to find different names for what are sometimes clearly specimens

belonging to the same species. Many of these color forms or varieties glow under black light, in particular those forms occurring at depths below 10 m (30 feet).

It appears that two or three genera are frequently available from aquarium shops: the smaller, often colonial *Actinodiscus* species and the larger, mostly solitary *Rhodactis* species (also sometimes referred to as "elephant ears"). The former have an average disc diameter of about 3 to 5 cm (1.2-2 inches) in adult specimens, while the latter grow to a respectable 20 cm (8 inches) and more across the oral disc. The third genus often listed is *Discosoma,* which is of uncertain validity.

Mushroom anemones make excellent inhabitants for the miniature aquarium. Because they rely extensively on the photosynthetic activity of intracellular algae, most

Mushroom anemones, mostly *Actinodiscus* species, have become very popular in marine aquariums, where they live very well under high-intensity lighting. Photo: U. E. Friese.

Most *Actinodiscus* species have very short tentacles arranged along regular rows much like a mushroom coral (*Heliofungia*). Colors vary tremendously, but brown types are perhaps most commonly kept. Photo: Dr. H. R. Axelrod.

need no or only little supplementary food, although in their natural habitat they are principally feeders on planktonic organisms. Their coloration can be dramatically affected by different light levels and intensities. *Actinodiscus* specimens of the same species and within the same colony often exhibit different shades of red, brown, and green. Fluorescent green and blue/violet animals that will glow under black light are common in this genus. Since mushroom anemones always have to be transferred together with the rocks or pieces of coral to which they are attached, they are ideal for constructing miniature reefs in a sufficiently large tank. In fact, they are often sold as "mushrooms on rocks" in aquarium shops.

Species most commonly traded are the reddish brown *Actinodiscus numiformis* (Indo-Pacific) and *Rhodactis sanctithomae* (Caribbean). Again, it must be emphasized here that there is much confusion about the specific identity of many forms. This point is best illustrated by the fact that a recent aquarium publication shows photographs of 18 *"Actinodiscus* species," eight *"Rhodactis* species," and several *"Discosoma* species."

Mushroom anemones are quite impossible to identify at the moment. Even preserved specimens yield different names when examined by different experts. None can be confidently identified to the species level, and there are major arguments as to whether the genus of most species should be *Actinodiscus* or *Rhodactis*. Photo: P. Wilkens.

Usually the larger specimens of mushroom anemones are assigned to the genus *Rhodactis* rather than *Actinodiscus*, but there is little scientific reason for this. This large brown mushroom anemone, *Rhodactis*, contrasts greatly in size with the small *Actinodiscus* at its side. Photo: U. E. Friese.

Not all mushroom anemones are brown. This clump of *Actinodiscus* features bright blue animals, while others may be red or green. Coloration apparently depends to some extent on light intensity and quality of food as well as genetics. Photo: U. E. Friese.

Order Actinaria (TRUE SEA ANEMONES)

The so-called "true" sea anemones are today divided into two suborders, the Protantheae and the Nynantheae. Of these only the latter are of interest to marine aquarists. The Nynantheae is made up of 35 families. The largest is the family Actiniidae, which contains the bulk of the species within 44 genera. There are more than 1,000 species of sea anemones in this order, which includes many of the popular aquarium species, but overall they make up only a fraction of all actinians. While some species are quite small and plain—and often extremely rare—others are quite large and extremely colorful and conspicuous.

Popular aquarist belief has it that the more colorful aquarium animals come from tropical waters. While this may hold true, for instance, for the majority of fishes, it is not so for sea anemones. The majority of colorful sea anemones—and there are many—come from temperate and cold seas. Admittedly, the largest species of sea anemones are always found in tropical oceans.

Sea anemones are popular with marine aquarists not only because of the graceful form and attractive coloration of these animals, but also due to the fact that many of them, if properly cared for, will live for a long time in aquariums and some may even reproduce quite prolifically. The operative term here is clearly "if properly cared for," and this really can only be done in a correctly set up and operated miniature reef aquarium. Now let us look in detail at some of the more popular "true" anemone species that are popular with marine aquarists and are periodically available in aquarium and pet shops.

SUBORDER PROTANTHEAE

This suborder includes a single family, Gonactiniidae. From an evolutionary point of view, the anemones in this group are the most primitive ones. There are only a few of them. Anatomically they are characterized by the presence of eight complete septa (which relates them closely to the corals) that are firmly attached to the pharynx. There is also an ectodermal longitudinal muscle layer. Filaments are absent from the septa.

Gonactinia prolifera (Crystal Anemone)

This small sea anemone, less than 1 cm (0.4 inch) in height, from the British and Norwegian coastlines of the North Sea is the most familiar example of a protantheaean anemone. The delicate body is almost transparent, with slight tinges of white and pink. It is usually found attached to mollusc shells at a depth of 60 to 80 m (180-240 feet). The mouth is surrounded by two rows of eight tentacles. This anemone likes to change its location quite frequently, moving about by alternately attaching the oral and basal discs; it even displays feeble swimming movements. Similar in appearance is a small group of sea anemones from the genus *Protanthea*.

SUBORDER NYNANTHEAE

The great majority of all sea anemone species belong in this group, which displays considerable diversity in behavior, coloration, morphology, and overall sizes. All sea anemones in this suborder are characterized by septa with filamentous attachments, strong longitudinal mesenteric (septal) muscles, a special sphincter muscle, and a multitude of cilia (tiny cells with long flagella) around the stomodeum and down the pharynx. There are always more than eight complete septa.

TRIBE BOLOCEROIDARIA

These (two families, four genera, and some ten species) are the most primitive nynantheaean anemones; they are best characterized by the genera *Bunodeopsis* and *Boloceroides*. Superficially, they resemble the Protantheae by lacking certain muscles. The septal arrangement, however, conforms to that found in all nynantheaean anemones.

Marine aquarists are unlikely to encounter these sea anemones. They

Facing page: The boloceroid anemone *Bolocera tuediae* comes from cold Arctic waters and is unlikely to be seen in the home aquarium. Photo: U. E. Friese.

are listed here in order to afford a better overview of the phylogenetic composition of the sea anemones in general. Nynantheaeans are relatively inconspicuous sea anemones from the tropical Indian and Pacific Oceans. Some species appear to form well-defined symbiotic relationships with certain crustaceans.

Bunodeopsis strumosa grows to a diameter of about 3 to 6 cm (1.2-2.6 inches) and a column length of about 4 cm (1.6 inches). It often is attached to sea grasses, but it also is found on rocks and cliff faces. There appear to be two color variants; one is completely white and the other yellowish with brownish dots. The anemone crabs of the Indo-Pacific, *Lybia,* carry a small anemone in each claw. Older literature identified this anemone as *Bunodeopsis prehensa,* but more recent literature states the anemone to actually be *Triactis producta,* a tiny species of the family Aliciidae.

TRIBE ABASILARIA (FAMILIES EDWARDSIIDAE, HALCAMPIDAE, HALCAMPOIDIDAE, ILYANTHIDAE)

Again this is a group of sea anemones that rarely are ever encountered by the marine aquarist in aquarium shops. This account is provided primarily for the benefit of those who do a lot of diving in temperate and cold waters, from Puget Sound and the Northeast Pacific to the northern Mediterranean Sea and the North Sea, where the alert and observant diver may possibly encounter some of these anemones.

Abasilarian anemones are never conspicuous. They are all burrowers that—as can be expected—lack the typical pedal disc and its related muscles. Instead, the aboral end is rounded off and cone-shaped. Some species have a weakly developed base resembling a modified pedal disc. These sea anemones, which range in column length from 2.5 cm

to 25 cm (1-10 inches) usually are buried up to their oral disc in soft mud or fine sand/silt. Their columns are elongate and slender; the typical nynantheaean sphincter is absent. When endangered they withdraw into the substrate with great speed, so the special closing mechanism of a sphincter muscle is not required.

Those aquarists who are lucky enough to find some of these sea anemones are best advised to keep them under conditions as recommended for tube anemones *(Cerianthus)*. They should be given a deep, soft (fine sand or mud) bottom substrate, subdued lighting, and no turbulent water currents. Feeding does not appear to be a problem. As particle feeders these anemones will accept most fine foods.

The genera *Edwardsia* and *Halcampa* are typical representatives of this family series. The species in both genera are only about 2.5 to 10 cm (1-4 inches) long. *Peachia hastata,* on the other hand, may grow up to 25 cm (10 inches) in length. It has a flesh-colored or pinkish stem, few tentacles with reddish brown spots, and a gleaming white oral disc. This sea anemone occurs throughout the North Sea and extends southward to the northern Mediterranean Sea. Its habitat extends from the intertidal zone down to a depth of 220 m (720 feet), where it feeds primarily on planktonic organisms and fish larvae.

The cold water regions of the North Pacific and North Atlantic and adjacent seas are the home for *Edwardsia* species. In the North Sea we find *Edwardsia andresi* (up to 10 cm, 4 inches, long, green with yellowish brownish dots) and *E. longicornis* (5 cm, 2 inches, long, orange) on soft, muddy bottoms from just below the low-water mark to a depth of 700 m (2300 feet). In the same geographical region (high northern latitudes) occurs *Halcampoides purpurea*. This is a rather

attractive species, 10 cm (4 inches) tall, red in color.

In the mud flats along the North American West Coast, we find *Edwardsiella californica* in abundance. This barely 5-cm-long (2 inches), worm-like sea anemone has a slightly transparent body with eight longitudinal bands. Along the East Coast of North America there are several *Edwardsia* species, including *E. leidyi* from Vineyard Sound southward, *E. lineata* and *E. sipunculoides* from Cape Cod northward, and the 15-cm-long (6 inches) *E. elegans* from Eastport, Maine, southward to Cape Cod. *Halcampa duodecimcirrata* is a burrowing sea anemone that occupies crevices in the sea bottom and among rocks and occurs from the Bay of Fundy to Eastport, Maine. It moves about in a worm-like fashion.

Peachia hastata of the northern Atlantic is the only abasilarian anemone even occasionally available to aquarists. Photo: Dr. T. E. Thompson.

TRIBE ENDOMYARIA (FAMILIES ACTINIIDAE, ALICIIDAE, PHYLLACTIDAE, BUNODIDAE, STICHODACTYLIDAE, THALASSIANTHIDAE, MINYADIDAE, AURELIANIDAE, PHYMANTHIDAE, ACTINODENDRIDAE)

All species within this tribe have a well-developed pedal disc and a strong mesentarial (septal) musculature, plus a sphincter in order to be able to close off the inverted animal. The most representative group in temperate oceans is the family Actiniidae. Below are some of the species periodically available commercially to marine aquarists.

Phymanthus crucifer, an odd anemone with "warts" on the disc. Some workers put this species in the genus *Epicystis.* Photo: P. Colin.

One of the more widely distributed sea anemones is *Actinia equina*, the crimson sea anemone. As usual, this is a poorly understood species confused with similar species such as *Actinia tenebrosa*. Photo: Takemura & Susuki.

Actinia equina (Crimson Sea Anemone)

This sea anemone comes in a range of colors from red to purple, with various intermediate shades. It has an enormous geographical distribution extending from the northern Mediterranean Sea and along the southern African coastline to central Japan and eastern Australia. In view of such a vast distribution it is not surprising to learn that the relevant literature lists several geographical and ecological subspecies (for instance, *A. e. pontica* and *A. e. japonica),* as well as eight major color variants. All give birth to live young. Maximum size of a well-fed adult specimen is about 12 cm (5 inches) in diameter.

It is typically an intertidal, rocky shores anemone that often is exposed high above receding tidal waters, but invariably is found in small cracks and crevices that retain some residual water. Maximum abundance is around the low-water mark, where it lives in areas protected from wave and surf action. It has 192 tentacles arranged in six tentacular rings on the oral disc.

The most intense crimson red specimens are found in the Mediterranean Sea. This red seems to be diet-related. Specimens given only fresh fish flesh will eventually turn very pale; however, a diet supplemented with tubifex and small crustaceans clearly enhances the redness. This species is very responsive to chemical (food!) stimuli and sudden water current flow.

The intensity of coloration of *Actinia equina* and many other anemones depends to some extent on food. Crustaceans contain carotenes that promote red coloration. Photo: Tierfreunde.

A very pale specimen of *Actinia tenebrosa*, the Australian version of *Actinia equina*. This cold-water species is easily kept in an appropriate aquarium. Photo: U. E. Friese.

A. equina is regularly available through the aquarium trade and makes an excellent addition to a miniature reef aquarium for temperate water species. As an intertidal species *A. equina* can live very well in maximum temperatures of around 23°C (73°F) or even slightly higher (depending on the origin of the particular specimens).

Actinia equina makes an excellent marine aquarium inhabitant, and all its subspecies and color variants appear to do equally well in captivity. Moreover, specimens that are kept in warm sea water (about 23-25°C, 73-77°F) are often accepted by the clownfish *Amphiprion xanthurus* as a substitute for the giant tropical stichodactyline sea anemones.

Actinia tenebrosa **(Waratah Sea Anemone)**

This species, from the East Coast of Australia, is virtually indistinguishable in coloration and size from *A. equina*. In fact, it may well be a subspecies. It also is found under identical conditions as the previously discussed species. If kept under appropriate conditions, *A. tenebrosa* will breed quite readily in an aquarium. It gives birth to live young, and miniature sea anemones will often start to appear, initially along the bottom substrate and soon thereafter on the tank walls and decoration. Small *A. tenebrosa* are primarily planktonic feeders, but captive specimens will already take relatively larger pieces of food such as minced raw fish or shrimp and crab meat.

***Tealia coriacea* (= *Tealia felina*) (Tubular or Beaded Sea Anemone)**

Another widely distributed group of

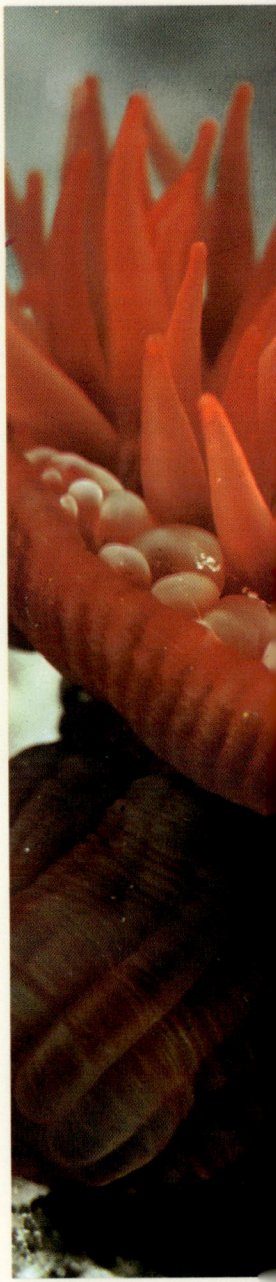

The stunning beauty of a fully colored Waratah anemone, *Actinia tenebrosa*, is hard to match. This is one of the few anemones that breeds regularly in the aquarium, producing living young. Photo: K. Gillett.

endomyarian sea anemones falling into the genus *Tealia*. They usually are stout and compact anemones with up to 160 tubular tentacles. The oral disc in some of the larger types ("species") can measure up to 30 cm (12 inches) in diameter. All species display considerable color variability. *Tealia* anemones show patches of red, yellow, green, and brown (and combinations of these colors). In addition, striped patterns are common. There rarely ever are two specimens alike within the same species. It is this variability in appearance that makes identifying individual *Tealia* species so difficult. Consequently, it is not surprising to learn

Tealia columbiana is one of the most familiar of California and East Pacific anemones, though cold water is required to maintain it. The photo on the facing page shows the characteristic swollen vesicles (verrucae) on the column. Photo at right: D. Gotshall; facing page: U. E. Friese.

that there is much confusion about the taxonomic legitimacy of certain "species."

In northern temperate and cold water regions *Tealia* usually is the most abundant type of sea anemone. In some areas, particularly those with protected rocky foreshores, there may be huge colonies of them. Basically, *Tealia* seems to prefer habitats just below the lowest water mark, although extreme low tides may expose some of these anemones on harbor pilings and other (normally) submarine structures. Most specimens are found close to wave and surf action (i.e., open ocean coast), but there they are always in

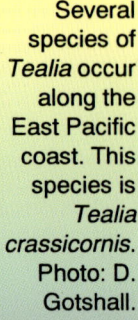

Several species of *Tealia* occur along the East Pacific coast. This species is *Tealia crassicornis*. Photo: D. Gotshall.

Tealia coriacea is just one of several northern, cold-water anemones of the genus *Tealia* that do fairly well in cool aquariums. Photo: U. E. Friese.

locations that are protected against direct wave impact (under or behind large boulders, etc.).

All *Tealia* species seem to do well in captivity; in fact, under optimum conditions they are known to have lived in aquaria for many years. They are typical predatory anemones that will accept all sort of foods, from earthworms to pieces of raw fish fillets, shrimp, clams, mussels, etc. DO NOT OVERFEED! In their natural habitat *Tealia* anemones are found under conditions of low or subdued lighting. This should be duplicated in the home aquarium, but the artificial tank illumination must include a light source that produces wavelengths with maximum depth penetration (actinic lighting). Symbiotic algae (zooxanthellae) do not appear to be present in any *Tealia* species. In captivity the water temperature must not exceed 15°C (59°F) for North Atlantic species, and 10°C (50°F) is the maximum temperature tolerated by *Tealia* of the North Pacific species (those from the Washington coast northward).

The column of most species has scattered wart-like bumps ("beads") of pinhead size to 7 mm (0.3 inch) in diameter that are of a white or similar very light color. It is common to find specimens that have broken shell fragments and other

Closely related to *Tealia coriacea* (and perhaps belonging to the same species) is *Tealia lofotensis* from the East Pacific. Photo: D. Gotshall.

138

Tealia felina probably is a synonym of *T. coriacea*. Notice the similarity in proportions to the *T. lofotensis* on the facing page, also probably a form of *T. coriacea*. Photo: A. Holtmann.

debris attached to their column.

European aquarium literature contains many accounts of *T. felina* being kept successfully in marine aquariums. This species is common in the North Atlantic, North Sea, and Baltic Sea, from shallow inshore waters down to considerable depths (100 m, 330 feet). Adult specimens have an oral disc diameter of about 7 cm (almost 3 inches) and a column height of about 5 cm (2 inches). The oral disc carries from 80 to 160 thick and blunt tentacles that normally are up to 3 cm (1.2 inches) long. They are arranged in two circles around the oral opening. This sea anemone is actually eaten by local residents in some of the northern European countries!

Along the West Coast of North America, *T. coriacea* is the most common

Tealia. There is some evidence that suggests that *T. coriacea* and *T. felina* are identical, although the body dimensions for the former are generally listed as being considerably larger than the latter. *T. coriacea* is stout and squat in shape, about 12 to 15 cm (5-6 inches) tall, with an oral disc diameter of about the same size. Beachcombers along northern California, Washington, and British Columbia shores commonly find it in rocky areas, sometimes exposed at extremely low tides, often in very narrow crevices. The predominant body color is a uniform scarlet or unrelieved olive brown, often with a blotchy color pattern. Radiating white lines among the tentacles are common. The tentacles are relatively thick and sharply tapering.

T. coriacea lives mainly in semiprotected waters, often partially buried in coarse sand (attached to a rock below the substrate surface), shell debris, or gravel, with only the oral disc with its tentacles showing. Particles from the surrounding substrate often remain attached to the anemone's column.

Another attractive *Tealia* from the northeastern Pacific Ocean is *T. lofotensis (= T. coriacea??)*. The relevant literature recognizes this species on the basis of its scarlet column with many white to pink tubercles arranged in longitudinal rows along the upper half of the column. There are four rows of tentacles, with each tentacle often banded red and white. There is some evidence that this is merely a variant or subspecies of *T coriacea*. The same can be said for the giant *T. columbiana,* which may be 20 cm (8 inches) across the oral disc and at least 25 cm (10 inches) tall. It should be pointed out here that in spite of some superficial similarities there also are characteristics that are

Tealia piscivora is another colorful East Pacific anemone. It is found on rocky bottoms from Alaska to the Gulf of California. Photo: D. J. Wrobel.

quite dissimilar between these forms, most notably the shape and length of the tentacles, which appear to vary widely.

All *Tealia* anemones make excellent marine aquarium specimens, but they must be kept in low water temperatures. It is important to ascertain where a particular specimen was collected. For instance, the temperature may be as high as 15°C (59°F) for some of the European specimens, and must be as low as 8-10°C (46-50°F) for Northeast Pacific *Tealia*. As true predatory anemones, they should be fed small pieces of raw fish, shrimp, clams, mussels, crabs, etc., at regular intervals (about two to three times per week, at the most).

Anemonia sulcata (European Wax Anemone)

This is one of the most abundant species in the eastern Atlantic (from West Africa northward to the northern British Isles) and the northern and eastern Mediterranean. Very characteristic for this species are numerous long and fairly thick tentacles that are non-retractile and have strong nematocyst action. Up to 200 tentacles have been counted in one specimen. They usually are yellowish white or grayish greenish with purple or lilac tips, rarely uniformly white. The oral disc is dark with radiating white stripes. The overall coloration appears to be dependent upon light intensity (380 nm actinic and 650 to 700 nm), and in fact *A. sulcata* must have adequate light levels otherwise it will die. Sexual reproduction is by means of releasing eggs and sperm into the surrounding water. A free-swimming planula larva will develop that eventually settles out as a small anemone.

A. sulcata is most commonly found along rocky shores in protected, sunny bays (often in massive colonies) from just below the low-water mark down to a depth of about 6 m (18 feet). As with most sea

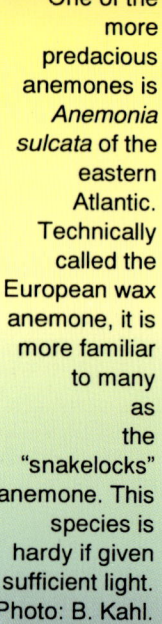

One of the more predacious anemones is *Anemonia sulcata* of the eastern Atlantic. Technically called the European wax anemone, it is more familiar to many as the "snakelocks" anemone. This species is hardy if given sufficient light. Photo: B. Kahl.

anemones, the largest specimens always are found in the deepest water.

Unlike most other anemones, this species can tolerate less than optimum water conditions. It often is seen to be alive and reasonably healthy in polluted coastal waters. It generally does well in a home aquarium, and in fact it often is commercially available from aquarium shops, especially in Europe. *A. sulcata* is a strong predator and will accept a wide range of food (raw fish, shrimp, beef heart, crab meat, mussel flesh, tubifex worms, earthworms, etc.).

A. sulcata forms close symbiotic relationships with a goby, *Gobius bucchichi,* the spider crab *Macropodia rostrata,* and a shrimp, *Periclimenes scriptus*. Because of the powerful nematocyst activity of this species, it is not recommended for community marine tanks.

Condylactis aurantiaca (Golden Anemone)

Although not particularly common over its natural range (eastern North Atlantic, northern and eastern Mediterranean Sea), it is frequently accessible to marine aquarists through the aquarium trade or by personal diving collections during vacation trips. Often it is found on a rocky bottom at the base of large boulders or rocky cliffs, partially submerged in sand and other fine debris (attached to a rocky surface underneath the sand).

The pedal disc usually is orange to scarlet, and the

In the Atlantic, the giant anemones often are species of *Condylactis*. The golden anemone, *Condylactis aurantiaca*, is one of the largest Atlantic species at 40 cm in disc diameter Photo: H Steiner.

143

This anemone often is identified in hobby literature as *Condylactis passiflora*, but it appears to be very similar or identical to specimens identified as *Condylactis gigantea*. Photo: U. E. Friese.

column has longitudinal white stripes and tubercles only along the upper third. The oral disc is grayish green with whitish tentacles, and the oral opening has a violet margin and is surrounded by four rows of golden yellow to ivory-colored tentacles. Some specimens may be shades of light brown.

This is a very large sea anemone that can reach a disc diameter of about 40 cm (16 inches). It inhabits relatively shallow water from just below the low-water mark down to a depth of about 6 m (18 feet). Since it lives partially buried, this must be taken into account when setting up the tank; a deep bottom layer of fine sand covering some flat rocks must be provided. It is very sensitive to damage to the pedal disc!

The eggs are fertilized in the body cavity of the female. They are released at the 12-tentacle stage. These young anemones can

Although the small *Gobiodon rivulatus* is more at home among coral polyps on the reef, in this aquarium it has accepted a zoanthid colony as home. Photo: A. Norman.

Two specimens, perhaps derived by fission, of the pink anemone *Condylactis passiflora*. Photo: U. E. Friese.

then easily be raised with *Artemia* nauplii or other live planktonic food.

This is a predatory sea anemone that will feed on a great variety of foods (same as for *Anemonia sulcata*), but it shows a strong preference for mussel flesh. Do not overfeed!

Condylactis gigantea (Pink-tipped Anemone)

This very large sea anemone is one of the largest species in the Caribbean, with a distribution from Brazil in the south to southern Florida (including the Bahamas and Bermuda) in the north. It is found both in shallow reef lagoons as well as along the leeward side of reefs. While it is fairly common, it never occurs in large numbers in any one area. *C. gigantea* usually is found from the low-water mark down to a depth of about 30 m (100 feet).

Species of *Condylactis*, such as this *C. gigantea*, usually expose only the disc and tentacles. The column and foot are wedged between coral blocks or buried in the substrate to help protect the animal. Photo: P. Colin.

The tentacles, in two or three rows, are long (often in excess of 10 cm, 4 inches) and pointed and usually are colorful but are highly variable. Juvenile specimens are particularly attractive, with tentacles that can be greenish, yellowish, or even pinkish with pink to violet tips. The oral disc may be with or without radial stripes.

Juveniles (up to 10 cm, 4 inches, in diameter) should be given preference for a community tank; large specimens may adversely affect other tank inhabitants (other sea anemones and corals in particular). This anemone is not accepted by clownfish *(Amphiprion),* but symbiotic relationships with commensal shrimp

The colorful anemone hobbyists call *Condylactis passiflora* is one of the more frequently seen sea anemones and is simple to maintain.

species such as *Periclimenes pedersoni* and *Thor amboinensis* are well documented.

Anthopleura xanthogrammica (Giant Green Anemone)

This is, beyond a doubt, one of the most attractive cold-water sea anemones, and it also is one of the most distinctive. Its vivid green coloration is due to the presence of enormous numbers of symbiotic algae (zooxanthellae) within its tissue. Specimens that are kept under subdued lighting usually turn white, but they seem to be able to live even without the symbiotic algae. In fact, captive specimens sometimes lose their vivid green coloration and turn into a dirty grayish green or even snow white.

A. xanthogrammica is found along the West Coast of the United States,

Although the coloration of *Condylactis passiflora* is quite variable, most specimens have whitish tentacles with bright pink tips. Photo: U. E. Friese.

from the Gulf of Alaska southward to central California. It appears that specimens from the southern end of its range are less intensely green than those from the more northern regions. It is a large anemone that can reach an oral disc diameter of up to 20 or 25 cm (8-10 inches). *A. xanthogrammica* inhabits rocky shores, where it prefers those sites exposed to considerable wave and surf action. Nevertheless, specimens kept in home aquaria normally do quite well, as long as optimum water conditions are being provided. The essential factors are abundant light (actinic lighting) and sufficiently low temperatures (preferred range 8-12°C, 46-54°F, depending on origin of the specimens). For feeding details refer to *Anemonia sulcata*.

Anthopleura elegantissima (Rough Sea Anemone)

This is a fairly small sea anemone with a disc diameter of about 5 cm (2 inches) in adult specimens. It lives along

Anthopleura japonica, a Japanese green anemone. Photo: Takemura & Susuki.

the West Coast of North America, from northern British Columbia southward to central California. It is a typical intertidal to upper tidal sea anemone. The center of abundance seems to be Puget Sound, where large, patch-like colonies can be found crowded together over rocks or in cracks and crevices, exposed at low tide.

A. elegantissima is characterized by conspicuous pink tentacle tips and its habit of crowding together to form clusters over large rocks and boulders, particularly those that are interspersed over sandy or muddy tidal flats.

Most specimens will have sand grains, shell fragments, and other debris adhering to the column wall. There is a three-fold reason for this. First, it shades the body against the sunlight (remember that this species often is exposed at low tide). Second, capillary action assures a continuous water supply along the body wall and so prevents desiccation. Third, the debris provides excellent camouflage.

Symbiotic algae in the tissue give this anemone an olive green color. Specimens kept in aquaria under less than optimum light conditions will usually turn flesh-colored or even white with pink tentacles.

A. elegantissima lives well in captivity, and small colonies can be maintained for many years. It will accept a wide variety of food, but food particles should be sufficiently small. Reproduction is

The colorful *Anthopleura midori* of Japan. One of the more attractive anemones, it unfortunately seldom is available. "Midori" means green in Japanese. Photo: Takemura & Susuki.

151

Many genera of sea anemones are never seen in shops. Some might make excellent aquarium animals and are collected locally by interested aquarists. Shown here is the South African *Pseudactina flagellifera*. Photo: Dr. T. E. Thompson.

primarily by budding, where young sea anemones appear at the base of adult specimens, starting to grow out onto the rock or inside the crevice, thus continuously enlarging the colony.

This is also a true coldwater sea anemone, and the water temperature must never exceed 12°C (54°F) (ideal range 8-10°C, 46-50°F). Clearly, specimens kept in home aquaria will, for the most part, need to be kept refrigerated. There are a number of excellent refrigeration systems available at reasonable cost. Those aquarists with an inclination toward tinkering can easily assemble their own refrigeration unit from components salvaged or cannibalized from discarded water drinking fountains or old soft-drink vending machines.

From about the same geographical region and similar habitat as *A. elegantissima* comes the slightly more fragile *A. artemisia*. This is essentially a burrowing sea anemone, attaching itself to a rock below the muddy or sandy bottom so that the tentacles lie flat against the substrate. The exposed tentacles are long and slender, sometimes semi-transparent but often mottled with brown or gray against a dark greenish body color.

Aquarists who want to collect this species will have to do a bit of digging. The body of *A. artemisia* is long and slender, and the basal disc generally is curved around a submerged stone that anchors the anemone firmly into the ground. Usually it is found below the low-water mark, on beaches where sand is

Anthopleura xantho-grammica is the common giant green anemone of the East Pacific. An abundant and hardy cold-water species, it needs high light intensities to support its zooxanthellae. Photo: U. E. Friese.

If the light intensity is lacking, the zooxanthellae of *Anthopleura xanthogrammica* die, producing a bleaching effect. Notice the whitish to pale green anemones in this photo Photo: K. Lucas, Steinhart Aquarium.

intermixed with rocks or along a rocky reef. For aquarium maintenance details refer to *A. elegantissima*.

Marine aquarists in Europe often keep the olive-green and white-spotted *A. thallia* and the red-spotted *A. ballii*; both species occur along rocky shores of the eastern North Atlantic and in sections of the northern Mediterranean Sea. *Anthopleura (= Aulactinia) crassa* is a spectacular yellow sea anemone with red nodules along its column. It occurs in northern regions of the Mediterranean Sea, but it is not common. This sea anemone is much sought after by European marine aquarists.

Oulactis muscosa (Australian Speckled Anemone)

This is a sea anemone found on rocky foreshores

along eastern Australia. It is popular with Australian marine aquarists. Common on the exposed open coast as well as in protected bays (Sydney Harbor), it can be found at any tide level, but always under water in shallow pools and cracks and crevices. In narrow cracks there may be a number of these anemones strung out for a meter (3 feet) or more.

This species is well camouflaged by attaching bits and pieces of debris, shell fragments, sand grains, etc., to its column and also distributing them among the tentacles. Since most of the body is hidden

Oulactis muscosa of southern Australia is easily collected locally and thus popular with Australian aquarists. Photo: U. E. Friese.

The column of *Oulactis muscosa* is heavily camouflaged with bits of shell, worm tubes, pebbles, and other debris. Photo: U. E. Friese.

Facing page: Marine aquarists with access to a productive shoreline often try to collect their own anemones, though this can be dangerous to the anemone (the pedal discs often are hard to remove, and even a small cut can lead to bacterial infections) and probably is illegal in many areas. Always check your local laws before collecting anything. Photo of *Oulactis muscosa*: U. E. Friese.

and only the outer portions of the tentacles normally are visible, it is difficult to determine the actual color of this anemone. In order to take a specimen for a home aquarium it is important to select one that can be removed together with its base rock. A close-up inspection then reveals a pale to dark greenish gray body marked with dark red to olive-green tubercles. The tentacles are short, in three rows, with grayish white horizontal bands. There is a series of short, fringed tentacles alternating between the outer plain tentacles. These short tentacles are largely covered by sand grains and debris and are visible only upon close examination.

O. muscosa is an excellent sea anemone for a miniature reef

Facing page: Like many anemones, *Oulactis muscosa* hides its column deep within crevices whenever possible. Photo: K. Gillett.

aquarium because it is small (about 8 to 10 cm, 3-4 inches, across, maximum) and to some degree colonial. Moreover, it can live together quite well with other intertidal invertebrates from the same region and a few small fishes. Its nematocyst action is surprisingly powerful. It is a predatory sea anemone, feeding mostly on small crustaceans and fish larvae in the wild. In captivity it will accept a number of substitute foods, which should be given in small pieces once or twice a week. DO NOT OVERFEED!

The required water temperature is about 12 to 22°C (54-72°F). This range is due to seasonal ambient temperature variations, and it represents the temperatures recorded in Sydney Harbor. Specimens from the open coast usually live in water that is marginally colder, although tide pool temperatures during the summer often exceed this range temporarily by a considerable margin.

Cnidopus verater (Bottle-green Anemone)

This species generally is found in the same habitat as *Oulactis muscosa*, but it occurs around the entire southern (temperate) half of Australia. Although *C. verater* is common in tide pools, it lives only in those that retain a considerable amount of water. As indicated in the common name, this is a green sea anemone that can be bright green or a dark bottle-green. The latter color seems to be more common, but intermediate shades also exist.

Cnidopus has three rows of tentacles each about 4 cm (1.6 inches) long and always of uniform coloration and without additional markings. The column wall usually is clean (little debris or shell fragments) and covered with tiny tubercles (papillae). The oral disc in adult specimens can reach a diameter of 4 to 8 cm (1.6-2.2 inches).

C. verater also is an excellent aquarium

inhabitant. For maintenance details refer to *Oulactis muscosa*.

Lebrunia danae (Stinging Branched Anemone)

The members of the family Aliciidae are characterized by a maze of pseudotentacles that arise below the rim of the oral disc and effectively disguise the real tentacles. In fact, the entire sea anemone is virtually hidden from view. The diameter of the irregularly shaped disc can reach 30 cm (12 inches) in adult specimens. The pseudotentacles carry

Cnidopus verator, the bottle-green anemone, is a graceful cool-water species easy to maintain in an appropriate aquarium. Photo: U. E. Friese.

Like most anemones with greenish coloration, the shade varies considerably depending on lighting and probably on food and stress as well. Photo: U. E. Friese.

large, whitish, oval or circular nodules (= acrorhagi) that contain massive batteries of nematocysts capable of stinging humans.

This species has a typical Caribbean distribution, ranging from Brazil northward to the Bahamas and Bermuda. It lives well below the intertidal zone and is most commonly found at depths from 2 m down to 60 m (6.6-198 feet). Pederson's cleaning shrimp, *Periclimenes pedersoni,* sometimes is found associated with this sea anemone.

Lebrunia coralligens (the coral branched anemone) lives in the cracks and crevices of coral heads, so that a diver can see only the pseudotentacles, with their conspicuously enlarged, often double-lobed tips that frequently bear a white circular band. The distribution is similar to that of *L. danae,* except

This young *Lebrunia danae* shows the greatly branched pseudo-tentacles. Photo: P. Colin.

Top: Only the mass of branched pseudotentacles shows in this large *Lebrunia danae*. The column is deeply buried in the substrate. **Bottom:** The pseudotentacles of *Lebrunia coralligens* are larger than those of *L. danae* and have very large rounded tips. Photos: P. Colin.

This photo of *Phymanthus crucifer* shows the bases of the tentacles and the disc to be covered with granules or vesicles that are the source of the common name beaded anemone. In some specimens the granules extend outward to cover the entire tentacles. Photo: P. Colin.

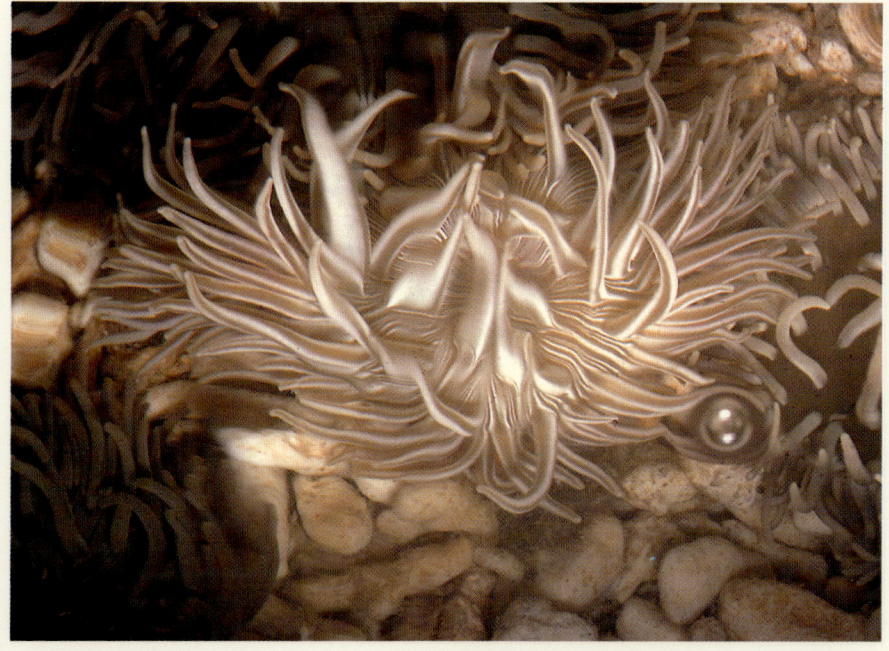

This attractively striped *Phymanthus crucifer* represents only one of several patterns found in the species. Some are banded and others may be uniformly greenish. Photo: U. E. Friese.

this species prefers shallower water (1 to 10 m, 3.3-33 feet).

Phymanthus crucifer (Beaded Anemone)

This anemone is a very attractive species that sometimes finds its way into American aquarium shops through imports from the Caribbean. It is characterized by an irregularly shaped oral disc up to 15 cm (6 inches) in diameter with a number of folds in its outer margin. There may be in excess of 200 tentacles that are striped, banded, or bear a granulated pattern.

The column usually is buried deeply in sand or in cracks and crevices, so that only the tentacular crown protrudes just above the substrate.

Dofleinia species (Banded Sea Anemone)

Dofleinia sea anemones inhabit the coastal waters of central and southern Japan (Hokkaido, northern Honshu). Most of the dark gray to brown column is buried in sand or mud, where the pedal disc is attached to a submerged rock or some other solid object.

This is a group of sea anemones characterized by extremely long, tubular tentacles that often are partially entwined and coiled in a disorderly arrangement on top of the oral disc; the oral opening generally is not visible (covered by the tentacles). In some species the tan tentacles have conspicuous narrow dark bands; others have uniform patterns of tiny dark dots or are dark olive-green. Pink tentacle tips are another frequent feature of *Dofleinia* species.

Dofleinia anemones are typical predators that will feed on a variety of foods (small pieces of fish, mussel, clam, and similar foods). The preferred temperature range appears to be 12° to 18°C (54-65°F).
CAUTION: Strong nematocyst action!

True cold-water anemones, the species of *Dofleinia* seldom are available to aquarists. Their simple tubular tentacles usually have pale spots or stripes that represent clusters of nematocysts. Photo: U. E. Friese.

A more attractive *Dofleinia* with pinkish tentacles showing the white spotting more clearly. Photo: U. E. Friese.

Phlyctenactis tuberculosa **(Australian Swimming Anemone)**

This is one of the few free-living and regularly swimming sea anemones. It occurs widely along the southern half of Australia. Its maximum column height is about 14 to 16 cm (6 inches), with a disc diameter of at least 12 cm (5 inches). Unlike most other anemones, it does NOT frequent a rocky habitat. Instead, it is intimately associated with algae pools and kelp beds.

In Sydney Harbor, this anemone usually is found among or loosely attached to the short brown kelp *Ecklonia radiata*. The fronds of *Ecklonia* strongly resemble the brown color and column nodules of the anemone and thus provide superb camouflage.

P. tuberculosa is characterized by the presence of a dense cover of bulging, bubble-like vesicles all along the column. The predominant colors of the column range from brown to reddish

One of the most unusual anemones that appears occasionally in the hobby is Phlyctenactis tuberculosa. The numerous inflated vesicles on the column aide the animal in floating in shallow water. Photo: U. E. Friese.

brown ("rust colored") and even shades of pink. The tentacles may be of a delicate light brown. This species has been kept for long periods of time in captivity at the Taronga Zoo Aquarium and Sydney Aquarium. Strong light is essential for the massive symbiotic algae in this anemone, and it seems to be able to survive for long periods of time without supplementary feeding. However, it clearly is a predatory anemone; it sometimes is found wrapped around baited fishing lines!

A very similar species is *Phlyctenanthus australis,* which lives primarily attached to rocky surfaces. The column is covered with bubble-like vesicles. Overall this anemone is grayish, brownish, or a delicate pink. It is found just below low-water mark around Southern Australia.

It is not unusual to see *Phlyctenactis tuberculosa* floating slowly along in shallow waters of the southern Australian coast. Photo: U. E. Friese.

Less frequently seen (at least by home aquarists) are the swimming sea anemones of the family Minyadidae, which are endomyarian sea anemones. They have a special adaptation, a unique swimming apparatus that is formed by a special morphological adaptation of the pedal disc. These anemones swim with their oral disc pointing downward, just below the surface. The genera *Minyas* and *Nautactis* occur in tropical seas of the Southern Hemisphere. It is questionable, though, whether these sea anemones retain their pelagic existence throughout their entire life. Sexually mature specimens have so far never been collected from among these "pelagic" forms.

The reader's attention is also drawn to the behavior of *Stomphia coccinea,* of the tribe Mesomyaria.

A group of glass anemones, *Aiptasia*. Unlike many familiar anemones, these cannot pull the tentacles down into the gastric cavity. Photo: U. E. Friese.

The southern Australian *Phlyctenanthus australis* closely resembles *Phlyctenactis* but has thinner vesicles and usually is found attached to rocks. Photo: U. E. Friese.

Beauty on the reef: clownfish (in this case *Amphiprion bicinctus*) in their anemone, *Entacmaea quadricolor*. Though taken in the Red Sea, this photo could be repeated with various species throughout the Indo-Pacific. Photo: G. Spies.

GIANT SEA ANEMONES (FAMILIES ACTINIIDAE, THALASSIANTHIDAE, STICHODACTYLIDAE)

These are the famous sea anemones that live symbiotically with various clownfishes of the family Pomacentridae. They are endomyarian anemones that come principally from three different families. The family Actiniidae contains two species known to be frequented commonly by clownfishes. Then there is the family Thalassianthidae, which has only one species commonly associated with clownfishes. The largest number of sea anemone species that have a strict symbiotic relationship with clownfishes belong in the Stichodactylidae; at least eight species from this family are known to host clownfishes.

Since these anemones are of particular interest to marine aquarists, they are discussed in detail later.

Mesomyarian and Acontian Anemones
TRIBE MESOMYARIA

The anemones in this tribe are very similar to those discussed in the tribe Endomyaria. The main distinguishing feature setting these two groups apart is in the location of the sphincter muscle. In the Mesomyaria it is situated in the mesoglea (the structureless layer between endoderm and ectoderm). Most of these

Facing page: Clownfishes (also called anemonefishes) belong in the genera *Amphiprion* and *Premnas* and are damselfishes, family Pomacentridae. Other species of damselfishes, especially of the genus *Dascyllus*, the humbugs, also occur in anemones. Photo of *Amphiprion clarkii*: P. Laboute.

Right: Tide pools in the Pacific Northwest are home to many anemones both colorful and dull.

anemones live in cold water, often at great depths, and are therefore essentially inaccessible to most marine aquarists, except those who are avid SCUBA divers in the cold waters of Puget Sound, for instance, where at least one species is fairly common. There are a total of three families, 23 genera, and 72 species.

Stomphia coccinea (Swimming Anemone)

This is a typical representative of the Mesomyaria, and possibly one of the few sometimes accessible to those marine aquarists who do their own collecting in the cold seas and oceans of the higher (northern) latitudes. *S. coccinea* is a circumpolar species that lives attached to rocks in the intertidal zone down to a depth of about 450 m (935 feet). The maximum column size is about 7.5 to 10 cm (3-4 inches). The normal habitat is a rocky or rubble-strewn bottom. Coloration varies

Some clownfishes have very large ranges (literally Africa to the South Pacific) and are quite variable in color. These *Amphiprion clarkii* from the Maldives are very dark compared to some other populations. Photo: Dr. H. R. Axelrod.

Facing page: Clownfishes have an unusual social structure in which females can become males when necessary. Photo of *Amphiprion chrysopterus* on *Stichodactyla giganteus*: Dr. G. R. Allen.

considerably. The column is cylindrical, cream to yellow with crimson spots. The whitish to yellow tentacles (in four rows) are semitransparent, with reddish encircling bands. The oral opening has a reddish margin.

Stomphia is known for its unique predator response—its ability to swim away from any potential enemy. It often is found by SCUBA diving aquarists in Puget Sound, Washington. Specimens from that area must be kept at water temperatures from 8 to 12°C (46-54°F).

Other mesomyarian sea anemones include the somewhat larger *Actinostola callosa* (from the North Atlantic and North Pacific) and *A. abyssorum* (from the northern coastline of Norway only). The latter may be merely a color variety of the former. Both species obtain a column height of about 25 cm (10 inches), with an oral disc diameter of about 20 to 25 cm (8-10 inches). They are essentially deep-water species. *A. callosa* occurs at depths from 40 m to 2000 m (130-6,600 feet). It is a rather attractive sea anemone with a whitish column wall that often has a tinge of red or blue. The similarly colored oral disc has dark radial lines, and the tentacles are dark brown.

TRIBE ACONTIARIA

The characteristic feature of the anemones in this group is the presence of acontia. When touched, the anemone will eject long, thread-like filaments through the oral opening or through pores in the column wall. These threads are about 0.5 mm (a fiftieth of an inch) thick and very poisonous. Any unwary fish that mistakes them for food will probably die. A sphincter muscle (to close off the body cavity with the tentacles withdrawn inside) may be absent, located in the endoderm (as in the Endomyaria), or (more frequently) located in the mesoderm.

This large group contains many popular aquarium species. They are contained in 11 families and 46 genera.

FAMILY HORMATHIIDAE

The sea anemones in this family are characterized by a compact, cartilage-like pedal disk that is quite hard. It is often covered by a cuticle, which usually is shed under aquarium conditions. The acontia are ejected through the oral opening; there generally are no pores in the column wall. Most of these anemones occur either in temperate or cold water or live at great depths. This family contains 15 genera and 103 species. Few of them are known to do well in captivity, except a few very interesting symbiotic forms that occur in shallow subtropical waters.

Calliactis species (Hermit Anemones)

This genus is widely represented in subtropical and tropical waters, where it usually lives at considerable depth. Some species inhabit temperate waters of the higher northern latitudes. Most sea anemones in this genus are remarkably similar in color and appearance. The body shows various shades of brown covered with irregular yellowish or whitish patches and stripes. The tentacles are whitish or yellowish to pink. The species reproduce by means of eggs released into the water, where they are fertilized by free-floating sperm.

Calliactis forms rather firm symbiotic relationships with marine snails of the genus *Murex* and a facultative symbiotic relationship with various hermit crabs that select *Murex* shells as their home. This symbiosis has been reported for *C. parasitica* (from the Central to North Atlantic and the northern Mediterranean Sea) and for *C. tricolor* (from

A hermit anemone, *Calliactis* species, attached to a rock. Though these anemones often are found attached to hermit crabs and their shells, they also are found living free. Photo: Dr. L. P. Zann.

northern Brazil northward to southern Florida and on to Bermuda). There appear to be a number of geographical "forms" (species ?) of *Calliactis* that all exhibit the same symbiotic behavior.

The aquarium literature reports that this genus must be kept below 20°C (68°F). This may apply to those species from more northern (higher) latitudes, but I have seen specimens from southern Florida and the Caribbean that did quite well at temperatures around 23 to 25°C (73-77°F). Feeding does not represent a problem. Small pieces of chopped raw fish, shrimp, mussels, etc., will be taken.

Adamsia species (Mantle Anemones)

This is another sea anemone known primarily for its well-defined symbiotic relationships with certain hermit crabs *(Pagurus)*. In fact, it is never found without its hermit crab partner. Should the hermit crab die, the sea anemone will succumb shortly thereafter. The term "mantle" anemone refers to the sea anemone's behavior of virtually encasing (like a mantle) the hermit crab's shell. Moreover, due to the anemone's position on the shell, part of the shell is dissolved and the living area for the crab is enlarged due to a horny secretion from the anemone's pedal disc.

This genus has a wide geographical distribution from the tropics to the temperate and cold waters of the higher (northern) latitudes. Again, there appears to be conflicting information in some of the aquarium literature. It is often stated that *Adamsia* cannot be kept at water temperatures in excess of 18°C (65°F), yet specimens from the Caribbean seem to do well in a typical tropical miniature reef aquarium. There may well be different species involved here. *A. palliata* is found commonly in the

This Mediterranean hermit crab, *Dardanus arrosor*, bears two large *Calliactis parasitica*. These anemones will do well in the aquarium (with or without their crab) if kept near the same temperature at which they were caught. Photo: U. E. Friese.

Central and North Atlantic, from the northern Mediterranean Sea to Norway. Certainly tropical species of *Adamsia* seem to do well in captivity, provided they are not separated from their symbiotic hermit crab partner.

Anthothoe albocincta (Little Striped Anemone)

This is a small acontiarian anemone that is found along the eastern and southeastern coasts of Australia. There it lives in often large colonies under rocky ledges, just at or below the low-water mark. It is easily recognized by its characteristic longitudinally striped column and the radial stripes on the oral disc. While this species may show some variations in its overall coloration, the stripes remain the particular feature. Usually the oral disc is of a light brown color, set off against cream-colored tentacles. In

Anthothoe albocincta is a small species with green or orange stripes on the column. It is common near jetties and in other shallow waters of southern Australia. Photo: U. E. Friese.

Facing page: When touched, *Anthothoe* reacts by releasing large numbers of nematocysts from small pores in the column. These can cause a burning rash in humans (upper photo) though usually not serious. Photos: W. Deas.

mature specimens, about 2.5 cm (1 inch) in diameter, the pattern of stripes becomes somewhat faded but still remains recognizable.

This species does well in captivity. Australian marine aquarists keep it at about 12 to 20°C (54-68°F) with strong seasonal variation. It feeds on small organic particles and zooplankton, but will readily accept brine shrimp larvae and even adult brine shrimp. This species ejects acontia through tiny holes along the column wall.

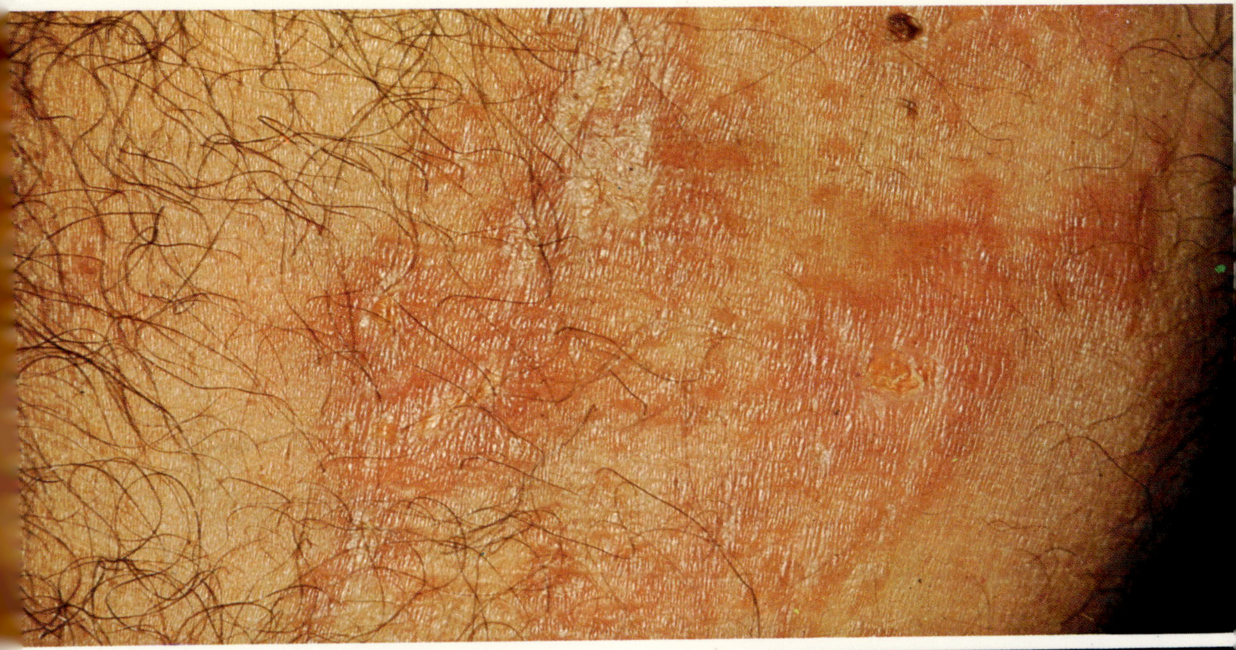

FAMILY SAGARTIIDAE (SAGARTIAN ANEMONES)

This is another group of sea anemones (ten genera and 62 species) that live predominantly in temperate and colder seas. Few of them are ever traded commercially, but European marine aquarists sometimes have access to a couple of species. On the other hand, SCUBA diving aquarists in the Northwest Pacific region (e.g., Puget Sound and adjacent waters) may occasionally come across one or two very attractive sagartian sea anemones.

The name given by Gosse to this family in 1855 has an interesting history. The Sagartians were a tribe of warriors under Xerxes, an ancient Persian king. These warriors were famous for using a sling to hurl ropes at their enemies, the horses and men becoming entangled in the ropes. The analogy to the sagartian anemones is fascinating: when touched or attacked, the anemones eject rope-like acontia through the oral opening or from pores in the body wall!

Sagartia rhododactylus (Delicate Anemone)

This small sea anemone, about 2 to 5 cm (0.8-2 inches) in diameter, occurs in the eastern North Atlantic on the shores of the Iberian Peninsula, France, England, and Scotland, to Norway and Iceland. It prefers rocky shores, but quite often specimens are found attached to kelp. The coloration is highly variable, and at one time five separate species and numerous color varieties were named in the literature. The prevalent coloration of this species (and apparently of its relatives) is pinkish to orange or brown, often with some paler vertical bands.

This attractive little anemone has very fine, delicate tentacles (up to 200) arranged in five concentric circles on the

Facing page: Though small and fragile in appearance, *Anthothoe* species should be respected. They have earned the name "stinging anemone" in Australia. Photo: K. Gillett.

Facing page: Exceptionally colorful plumed anemones may be bright orange if fed the proper foods, but usually they fade to a pale orangish white.

oral disc. They are orange, white, or pinkish to purple. The oral opening is either uniformly white or yellow, or brownish and pink. This species is open during the day, while *S. troglodytes* from the Mediterranean Sea remains closed during the day and is then mostly buried in sand. It only opens up at sunset and during the night. There is at least one species in Puget Sound and along the Washington and British Columbia coastlines. It is primarily semi-transparent pink with relatively long, thick, pointed tentacles.

For marine aquarists who are about to purchase a sagartian anemone, it is important to ascertain where a particular specimen came from. Invariably, these anemones must be kept at low water temperatures, about 15°C (59°F) maximum for animals from the North Atlantic and about 18°C (65°F) maximum for the Mediterranean species. Sagartian anemones from the Northeast Pacific (Puget Sound and adjacent waters) must be given a fairly uniform water temperature of 8 to 10°C (46-50°F).

FAMILY METRIIDAE (PLUME ANEMONES)
Metridium senile **(White-plumed or Piling Anemone)**

This is probably one of the most attractive sea anemones; it certainly is one of the most characteristic and conspicuous ones. *M. senile* is essentially a circumboreal species, and the relevant literature lists five different varieties. It occurs from the subpolar Arctic region southward into the North Atlantic as well as into the North Pacific. In the North Atlantic it is found as far south as northern Spain (but not in the northern Mediterranean Sea) and along the East Coast of the North American continent as far south as Delaware Bay. On the North Pacific side *M. senile* occurs as far south as central California

and down to northern Japan on the Asian side of the Pacific.

As the popular name indicates, this anemone looks like a large plume (the Europeans refer to the anemones in this family as sea carnations, a most appropriate name!). This "plumage" is created by the vast number of very fine tentacles that, in spite of their appearance, are not branched but are slightly curled, which gives this sea anemone its "fluffy" look. There are more than 1,000 tentacles!

Metridium can be up to 30 cm (12 inches) tall with a disc diameter of about 25 cm (10 inches). The deeper specimens occur, the larger are the individual animals. The most prevalent color is a snowy white, but orange, olive, brick red, and brown also are frequently seen. The orange form is, next to white, the most common

Do you see a resemblance to a carnation in this *Metridium marginatum*? Many people do. Photo: U. E. Friese.

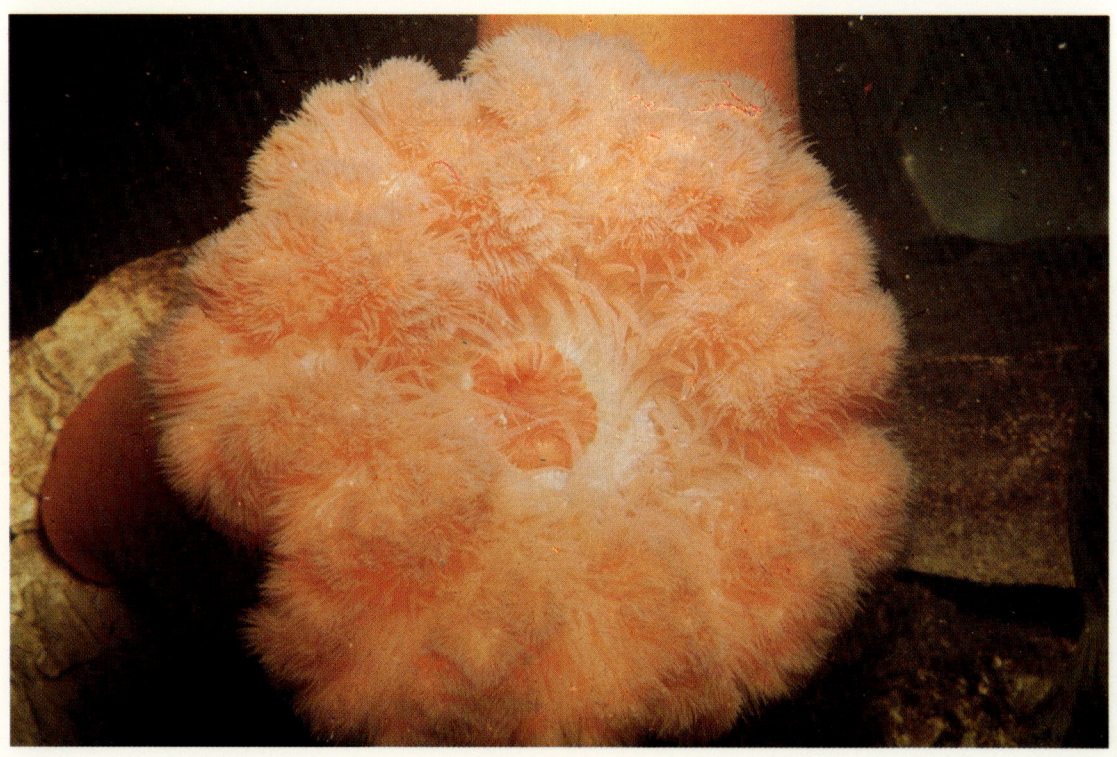

This little group of *Metridium senile* from Puget Sound shows the two main color forms (or perhaps species): stark white and orange. Photo: U. E. Friese.

form and sometimes is designated *M. marginatum*. The tentacular crown often is a shade lighter than the remainder of the body.

This anemone is commonly found in fairly shallow water, close to the surface. Aquarists are likely to find it on the underside of floating marinas and on the shaded side of pilings, wharves, docks, and on other submarine structures. For instance, in Elliot Bay (Port of Seattle) many small *M. senile* can be seen hanging limply from pier pilings at low tide.

M. senile prefers areas where there is rapid tidal flow but not areas of outright turbulence (wave action, surf). Specimens attached to wooden submarine structures can easily be removed with a scraper or putty knife. Those found on rocks should be transported together with the rock.

To be at its best, *Metridium senile* must be kept cool. This precludes successfully keeping it in most miniature reef aquariums. Photo: K. Lucas, Steinhart Aquarium.

Attempting to pry *M. senile* off a rock usually results in damage to the pedal disc, which tears easily. This anemone gives off a lot of slime and requires diligent handling plus initial maximum filtration as well as substantial water changes. In spite of these difficulties *M. senile* is popular with cold-water marine aquarists, simply because of its incredible beauty and delicate appearance.

Depending upon the location from which particular specimens have been collected, *M. senile* should not be kept at temperatures above 20°C (68°F). In fact, even at that level most specimens stop feeding and begin to degenerate. About 15°C (59°F) is a far more appropriate summer temperature for *M. senile*. During the winter months the water temperature can easily drop to 10°C (50°F). Specimens collected in the

The plumose appearance of *Metridium senile* results from multiple branching of the main tentacles plus numerous tiny secondary tentacles on each branch. Photo: D. Gotshall.

Facing page: Although called "glass anemones" in the literature, the few aiptasid anemones seen in the hobby are better described as "pedestal" anemones because of their shape. Large specimens often host clownfishes in the aquarium. Photo: Dr. H. R. Axelrod.

Northeast Pacific (off the Washington and British Columbia coastlines) prefer a temperature range from 8 to 12°C (46-54°F).

Metridium is a typical particle feeder and should preferably be given small planktonic food. *Artemia* larvae are a reasonably good substitute. A mixture of mussel flesh, shrimp, and some of the commercially available marine foods, together with a multi-vitamin mixture, could be finely macerated and passed through a blender.

It is best not to mix *Metridium* with other sea anemones in a home aquarium. *Metridium*'s strong nematocyst actions can severely affect other species. Reproduction is sexual, as well as asexual (by means of laceration).

FAMILY AIPTASIIDAE (GLASS ANEMONES)

This now brings us to the last group of anemones of interest to most marine aquarists, the glass anemones. Anatomically they are characterized by a longitudinal muscle band present only in the upper part of the column and only a very weak sphincter muscle; consequently, the tentacles can not be closed off in the gastric cavity. Reproduction is primarily by laceration and giving birth to live young. The young grow inside the gastric cavity and are relatively large at birth. There are three genera and about 20 species.

Glass anemones are of interest to marine aquarists primarily because sometimes they can appear in pest proportions. Quite often they are not deliberately purchased or collected, but are introduced as juveniles together with other animals. Living rock is often the carrier for the initial "brood stock" of aiptasiid anemones. Suddenly, there are small glass anemones all over the tank. While they are quite attractive and delicate looking, glass anemones can—when

occurring in large numbers—take over the tank to the detriment of other anemones. However, on the more favorable side they are excellent food for various large butterflyfishes (Chaetodontidae)!

The aquarium literature lists a number of species from different localities, but the species identifications in many cases appear to be doubtful at best. One of the more conspicuous species is the trumpet anemone, *Aiptasia couchii* (about 12 cm, 5 inches, tall), which occurs in the western Atlantic from West Africa and the Mediterranean Sea to northern England and Ireland. Its slender column often is trumpet-shaped and is crowned by up to 96 tentacles, which gives young specimens a gorgon-like appearance. The color of the column ranges from reddish to orange with white dots to almost bluish black. The oral disc usually has radiating stripes and is dark gray, slightly lighter around the oral opening.

Somewhat smaller (5 cm, 2 inches, tall) is *A. diaphana* from the Mediterranean Sea (including the Suez Canal). It often is referred to as THE glass anemone, since it is largely transparent.

One of the largest aiptasiids is the variable trumpet anemone, *A. mutabilis,* with a column height of about 20 cm (8 inches). It too comes from the Mediterranean Sea. The principal reason to find more species of glass anemones in the Mediterranean Sea than anywhere else seems to be the intensive biological-oceanographic research that has been going on for many years in that region through famous marine biological research stations at Naples, Monaco, and elsewhere.

One of the most prevalent glass anemones in the Caribbean is *Bartholomea annulata,* the ringed anemone, which is reported from throughout the Caribbean northward to the Bahamas and to Bermuda. It probably also

Glass anemones often develop from live rock and may appear in large numbers in a tank. They are eaten by some reef fishes, however, a point in their favor. Photo of *Aiptasia* sp.: Dr. H. R. Axelrod.

The small *Aiptasia tagetes* of the Caribbean often appears suddenly in tanks. Its small size and unimpressive appearance do not aid its popularity. Photo: P. Colin.

extends southward over a similar range.

Another small species from the Caribbean is *Aiptasia tagetes*. The oral disc is bluish white or nearly colorless transparent, and there are up to 96 brownish tentacles. It occurs down to a depth of about 15 m (50 feet). *A. tagetes* lives in a wide variety of habitats and is a durable aquarium species. This is one of the glass anemones often introduced into home aquariums via planula larvae or as adult specimens attached to living rocks. *A. pallida* is a glass anemone common along the coastline of the southeastern United States and the northern Gulf of Mexico.

Subclass *Octocorallia* (SOFT CORALS, GORGONIANS, SEA PENS)

Order Pennatularia (Sea Pens)

Not many marine aquarists will ever keep sea pens in their miniature reef aquarium, except possibly those who do some of their own collecting while SCUBA diving. On the other hand, sea pens commonly are on display in

One of the more commonly seen Caribbean anemones is *Bartholomea annulata*, a fairly large (to 30 cm or 12 in) species that often is home to various shrimp. The tentacles are ringed with white clusters of nematocysts. Photo top: Dr. H. R. Axelrod; bottom: U. E. Friese.

large public aquariums. They are indeed spectacular animals, and any aquarist who has ever kept one or two sea pens in his tanks will have been awed by their beauty.

Although sea pens are not closely related to sea anemones, they are sufficiently close that it is easy to justify their inclusion in this book. Instead of the typically hexagonal internal arrangement of sea anemones, sea pens have an octagonal mesenteric arrangement. Moreover, rather than being a single, individual animal—as are all sea anemones—sea pens are made up of colonies of polyps.

A sea pen consists of an elongated body of a primary axial polyp and a multitude of tiny secondary polyps branching off laterally, giving this animal its "feather" appearance. The body of a sea pen is divided into a proximal stalk or

The brown color of *Bartholomea annulata* is due to an abundance of zooxanthellae, is this case bearing predominantly brown pigments rather than green. Photo: U. E. Friese.

peduncle and a distal rachis, the region of the multitude of secondary polyps. The overall appearance often is much like an old-fashioned quill pen, but some types have very divergent and bizarre colony shapes.

Unlike sea anemones, a sea pen is never solidly attached to a base or substrate. Instead, sea pens simply burrow into a soft, muddy bottom for the entire length of the bare peduncle. The animal accomplishes this by alternately expanding and contracting the end-bulb, which also anchors the animal in place.

Depending upon the length of the primary polyp, sea pens often can reach sizes of up to 1 m (3 feet) in length. The maximum size ever recorded for a sea pen was 2.7 m (9 feet). Sea pens can inflate themselves with water to give them their maximum size.

Sea pens come in a wide range of colors, but red, brown, white, yellow, and orange are the dominant

If the light intensity is too low, the zooxanthellae die, reducing the contrasting background coloration and making the bands (annulae, the source of the specific name) of nematocysts more conspicuous. Photo: U. E. Friese.

In nature the foot and column of *Bartholomea annulata* are hidden deep within a crevice in the bottom or even in a large dead shell. Only the tentacles are visible when the anemone is feeding. Photo: P. Colin.

colors. When they are touched at night, they may emit a bright light that spreads in waves over the entire exposed surface. Although a well-defined skeleton is absent, sea pens do have an internal calcified stalk that gives the animal some rigidity within its primary polyp. The tiny secondary polyps can only be seen when a sea pen is totally inflated. Usually this occurs during the hours of darkness; during the day the animal is mostly deflated, which makes it shrink to the size and appearance of a dead or dying plant lying on the bottom.

Sea pens are cosmopolitan cnidarians occurring in all temperate and cold seas of the Northern and Southern Hemispheres. There are some 13 families and at least 30 species. Many of these look superficially very similar, and generic and specific distinctions are often based on minute internal structures and arrangements.

Not only is it the general inaccessibility of sea pens that makes them rarities in home marine aquariums, but they also are quite difficult to keep. The main problem is food and feeding. Here we must remember that sea pens are made up of a multitude of tiny polyps that are only a fraction of the size of even the smallest sea anemone. Consequently, food has to be equally small. Sea pens are particle feeders, feeding principally on live plankton. Any SCUBA diver who has come across these animals in the wild will have noticed that sea pens are always in areas where there is maximum current flow, but without any wave or surf turbulence. This is related to their food requirements for enormous, continuous plankton supplies. This is very difficult, if not impossible, to duplicate in a home aquarium. If these animals are to be kept successfully in marine aquaria, large amounts of plankton will have to be artificially cultured. Brine shrimp larvae are taken but seem to be insufficient as a staple food.

European public aquaria often display sea pens from the northern Mediterranean Sea and the eastern North Atlantic. *Pennatula phosphorea,* the slender sea pen, is nearly cosmopolitan in all cold seas. It grows to a height (when fully expanded) of about 40 cm (16 inches). The prevalent colors are dark red to white, and the polyps are white. As indicated by the common name, this sea feather is relatively slender in its build. It lives on sandy or muddy bottoms, but in captivity it often refuses to dig into the substrate, remaining lying open on the bottom. It is highly phosphorescent at night. Similar in appearance is the reddish *P. aculeata,* also from the northern Atlantic.

Quite a bit plumper is *Pteroides spinosum,* the plump sea pen. This animal reaches a maximum height of about

Sea pens, such as *Cavernularia obesa*, are not common in home aquaria. They are delicate, often require deep substrates in which to anchor successfully, and may need tremendous amounts of planktonic foods. Photo: U. E. Friese.

60 cm (24 inches), and its range extends from the northern Mediterranean Sea to the central North Sea coastline. It displays a highly variable coloration from brown to violet or even orange. The polyps are white or brown. A strong muscular foot enables this sea pen to burrow quickly.

Very common along the coast of Washington and British Columbia is the orange sea pen, *Ptilosarcus gurneyi,* which is sometimes found intertidally, but more often below the low-water mark on sandy bottoms. The upper part of the bilateral animals contains many tiny polyps of bright yellow-orange color. Also from the northeast Pacific comes the slender sea pen or thin sea whip, *Stylatula elongata.*

In some sea pens, bilateral symmetry is replaced with radial symmetry, as for instance in *Veretillum cynorium* from the western Atlantic and *Cavernularia obesa* from eastern Australia. In these species the rachis is a stout, radially symmetrical cylinder or it is club-shaped, and its entire surface is covered with relatively large secondary polyps. These particular species are considered to be of a more primitive stage of evolutionary development among sea pens.

The East Pacific sea pen *Stylatula elongata* in its natural habitat off California. Photo: D. Gotshall.

To successfully keep almost any cnidarian, whether a stony coral, an anemone, or a soft coral, you need high-intensity lighting that will allow zooxanthellae to survive and reproduce. Only pale, burrowing cnidarians, such as most sea pens, may actually be hurt by excessive light. Photo: P. Wilkens.

Symbiosis

Sea anemones are well known for intricate and well-defined relationships (= symbiosis) with other invertebrates—most notably crustaceans—and with certain fish species. Many of these relationships can be observed in a properly set up marine aquarium.

The term symbiosis is usually defined as the "living together" of two dissimilar organisms.

Facing page: The orange sea pen, *Ptilosarcus gurneyi*, occasionally is present in marine aquariums and is a rather sensational animal. The large bulbs at the base serve to anchor the animal in the bottom and help it burrow. Photo: Dr. T. E. Thompson.

Depending upon the degree and the nature of such an arrangement, further subcategories can be distinguished. Where both partners of a symbiotic relationship benefit, one speaks of mutualism. If only one of the two symbiotic partners seems to gain any benefit, with the other one being totally unaffected, the term commensalism is used. If, on the other hand, one species derives certain advantages to the obvious disadvantage or even harm of the other species, this is known as true parasitism. Sea anemones are well-known for rather intimate and mutually beneficial relationships with other animals.

A marked degree of interdependency can be observed between certain sea anemones and a few crustaceans. The sea anemone *Calliactis parasitica,* widely distributed throughout the central and northern Mediterranean (including the Adriatic) is known to attach itself to marine snail shells that often are inhabited by hermit crabs. To become established, the

The teddy bear crab, *Polydectus cupulifer*, usually carries small specimens of the anemone *Telmatactis decora* with it for protection. Photo: S. Johnson.

sea anemone first fastens its tentacle ring onto the shell, and then, by making a somersault, attaches its pedal disc to the shell. Several *Calliactis* specimens may be fastened to the same shell. Often this shell is occupied by the hermit crab *Pagurus bernhardus,* which will move on and look for a new "house" once it has outgrown the old one. Obviously this symbiotic relationship is a very weak one, since both symbionts appear to derive little direct benefit from it.

Markedly different is the arrangement between *Calliactis* and the hermit crab *Dardanus arrosor*. The crab carries only one sea anemone around with it on its shell. When the crab becomes too large and has to move into a larger shell, it will literally take the sea anemone with it. When a suitable new shell has been found by the

One of the most colorful Hawaiian crabs is the little *Lybia tessellata*. In addition to its odd color pattern, it is distinguished by often carrying juvenile anemones, usually *Triactis producta*, in its claws. Photo: B. Carlson.

211

hermit crab, it leaves the old one and climbs on top of the sea anemone. This causes the anemone to contract. With a little gentle tapping from the second and third walking legs of the crab, the anemone slowly opens up again. Then the crab actively pulls on the edge of the basal plate, which causes the anemone to detach itself from the shell. Once the *Calliactis* has come off the old shell, the hermit crab manipulates the sea anemone toward the new shell, where it will once again attach itself. If a *Calliactis* is removed forcibly under experimental conditions, the deserted hermit crab actually goes out looking for a new sea anemone symbiont.

An even closer symbiotic relationship exists between the sea anemone *Adamsia palliata* and the hermit crab *Pagurus prideauxi*. Neither of these two so different animals, once they have reached maturity, has ever been

The clownfish *Amphiprion allardi* in the giant anemone *Stichodactyla haddoni*. Notice the numerous tube anemones in the background. Photo: Dr. D. Terver.

found alone. This sea anemone is attached to the hermit crab's shell in such a position that it can receive leftover food from the crab. A special secretion from the basal plate of the anemone adds to the outer layer of the crab's house, enlarging it continuously. Consequently, the crab is rarely forced to look for a new, larger shell.

Various shrimp species are also commonly found living quite harmoniously among sea anemones. Most prominent are the shrimp *Periclimenes brevicarpalis* and *P. holthuisi,* which occur widely throughout the Indo-Pacific. These shrimps are partially transparent, yet they also display an intricate pattern of colored dots and patches that match the appearance of sea anemone tentacle tips. Moreover, the crustaceans are able to break up their profile and contours by alternately spreading and contracting their tail structures or bending their entire body. Incidentally, *Periclimenes* species occur not only throughout the tropics but also in the temperate waters of the Mediterranean Sea, where there also are similar symbiotic relationships between them and local sea anemones (e.g., with *Condylactis aurantiaca* and *Anemonia sulcata).*

Symbiotic relationships between sea anemones and other animals appear to be more common in tropical waters. Small porcelain crabs (family Porcellanidae) often are found among the tentacles of some tropical sea anemones, most notably the stichodactylid anemones. In recent years these crabs frequently have come onto the tropical marine aquarium market, especially in imports from the tropical Pacific and Indian Oceans. Depending upon the origin of the shipments, the crabs usually involved are *Petrolisthes maculatus* and *Petrolisthes ohshimai,* both often now placed in the genus *Neopetrolisthes.*

The small, often nearly transparent shrimp of the genus *Periclimenes* are found commensally with many invertebrates but are especially common on various anemones. Here a shrimp (possibly a male *Periclimenes brevicarpalis*) sits on the tentacles of *Entacmaea quadricolor*. Photo: G. Spies.

More of a "transportation agreement" rather than a true symbiotic relationship appears to exist between juvenile pink-tipped anemones, *Condylactis gigantea,* and the flame scallop, *Lima scabra.* Scallops are typical filter feeders, while *Condylactis,* equipped with strong tentacles, is a true predator. Consequently, the sea anemone seems to utilize the mollusc only for carrying it around and placing it into better contact with possible food sources. The powerful nematocysts of the sea anemone in turn provide excellent protection for the flame scallop against possible predators. Since both animals occur in abundance without each other's company, such joint arrangement appears to be more or less coincidental, rather than being indicative of a true symbiotic relationship.

Another type of symbiosis common among sea anemones involves their association with certain coral reef fishes of the genus *Amphiprion*. The large tropical sea anemones of the genera *Stichodactyla, Heteractis,* and *Entacmaea* (among others) are involved in these relationships. All species of *Amphiprion,* better known as clownfishes among marine aquarists, live in close contact with these anemones in their natural habitat.

In captivity, clownfishes also will accept other sea anemones as substitute hosts, even those that are relatively small. In general, this symbiotic relationship between fish and sea anemone takes place within strict and well-defined territories established by the fish. In fact, clownfishes (maximum length of about 10 cm, 4 inches), rarely venture far from "their" sea anemone. The usual swimming range of the fish is about 0.6 to 1 m (2-3 feet), seldom more than about 1.2 to 3 m (4-10 feet) away from the sea anemone. When

endangered and threatened, the clownfish will immediately dart into the tentacle crown of the sea anemone. Strangely enough, the powerful nematocysts of the sea anemone seem to have no effect upon the clownfish, which apparently can enter and leave the tentacles at will, while all other fishes will be attacked by the sea anemone.

Clownfishes use their sea anemones not only during the day as sanctuaries against predators in pursuit, but they also will swim into the tentacles of the sea anemone at night. Some clownfishes even lay their eggs among the anemone's tentacles. Despite such a close relationship there appears to be considerable discrimination on both sides. The sea anemone *Antheopsis* has an oral disc about 20 cm (8 inches) in diameter but will tolerate

If you look closely, you can see a small *Triactis* anemone in each claw of this little crab, *Lybia tessellata*. Photo: S. Johnson.

By laying their eggs at the base of a large anemone, in this case *Heteractis crispa*, a clownfish protects the nest from many possible predators. Photo of *Amphiprion chrysopterus* nest: Dr. G. R. Allen.

only juvenile clownfishes up to 5 cm (2 inches) in length. Larger fishes, even those forcefully removed from a larger *Antheopsis* and placed into a smaller one, inevitably will be attacked by the anemone. The same sea anemone will also attack smaller clownfishes if those have been removed from it only an hour earlier and are then returned to the same anemone.

This interesting phenomenon has received considerable attention from behavioral scientists. It has been established that the key to the protection afforded to the clownfishes from the nematocysts of symbiotic sea anemones lies in the fact that the fishes actually become acclimated. Initially the fishes are stung when they move into the sea anemone for the first time.

This hermit crab, *Dardanus*, is occupying a large tun shell bearing several anemones, probably *Calliactis* species. Over the millennia the crabs and anemones have developed a special relationship. When an anemone is transferred from one shell to another with the hermit, it is not simply ripped off the old shell. Instead, the hermit carefully coaxes the anemone to retract and gradually loosens the foot so transfer without injury becomes possible. Photo: K. Giwojna.

Eventually, however, the clownfishes are able to acquire an immunity against the stinging cells. When clownfishes come into direct contact with a prospective host anemone, the skin mucus of the fish becomes embedded with the substance inside the stinging cells of the anemone. This substance prevents the sea anemone's tentacles from attacking each other as well as impregnating the entire surroundings of the sea anemone. During this immunization or adaptation process the clownfishes become gradually immune. Their mouth region and the margins of abdominal and pelvic fins are protected first, then the entire throat region and the remainder of the body.

It has been suggested that clownfishes are able to protect themselves by somehow altering or affecting their own

The most commonly seen host–commensal relationship (at least in the aquarium) is that of a clownfish with its anemone. Photo of *Amphiprion allardi*: Dr. H. R. Axelrod.

In the aquarium a clownfish will accept many types of anemones not normally available on Indo-Pacific reefs. Photo of *Amphiprion akallopisos*: Dr. H. R. Axelrod.

protective mucous coatings (the typical fish slime that gives fishes their fishy smell). When this surface mucus is wiped off a clownfish that has been acclimated, it loses the protection and will be stung again by the sea anemone. This acquired immunity apparently is very specific between certain sea anemones and their respective fish symbionts. A fish immunized by one sea anemone is still quite susceptible to an attack by another sea anemone.

The presumed benefits to either one of the two symbionts (i.e., the sea anemone and the fish) have been under discussion in the scientific and lay literature since the middle

Facing page: Clownfishes can survive within the stinging cells of an anemone only through a period of gradual acclimation. To go from one anemone species to another requires time for acclimation or death may result.

of the 19th century. In recent years many marine aquarists have also grappled with this phenomenon, when both *Amphiprion* and the host anemones came to be imported in ever-increasing numbers.

The advantages to the fish are obvious. The powerful nematocyst-equipped tentacles of sea anemones give them excellent protection. Prolonged experiments and extended underwater observations in the wild have shown how totally dependent the clownfishes are upon their sea anemone hosts for their survival. *Amphiprion percula,* when taken about 10 m (33 feet) from its host, requires about 15 minutes to orient itself and find its way back to the sea anemone. Because of such a delay the fish provides an easy target for predators. Even small fishes will not hesitate to attack a lost clownfish as it struggles to find its way back.

A small school of *A. percula* captured, taken to the surface, and then released will immediately dart back down to the bottom. After a few minutes of obvious attempts to orient themselves, they will then swim toward their sea anemone. Should this host have been experimentally covered up or disguised, the fish will still swim in the correct direction, apparently using a well-developed spatial orientation system to find their way back.

As mentioned before, some of the large sea anemones also act as hosts for clownfishes at night. The fish then change their colors so that they blend in with the tentacles of the sea anemone. This adaptation makes it more difficult for nocturnal predators to find the fish among the tentacles.

The idea of tactile stimulation of the fish by the tentacles of the host sea anemone has received increasing support in recent years. When

clownfishes are taken into an aquarium without their particular sea anemone host, they will display bathing activity among clumps of algae just as they would among the tentacles of their regular host. The fish apparently also receive similar tactile stimulation from bathing in the stream of air bubbles emitted from an aquarium air stone or filter stem.

For many years it has been suggested that the sea anemones involved in a symbiotic relationship with *Amphiprion* are involved in the removal of certain parasites from the fish. Recent investigations, however, have shown that clownfishes living in sea anemones had a far heavier parasite load then did surrounding free-swimming fish species. Clearly, the sea anemones do not take part in the control of fish parasites.

Certain fish species actually feed on sea anemones, so the question arises: Do clownfishes also feed on their host anemones? Some researchers (as well as a few aquarists) have seen *Amphiprion* nibble on the tentacles of their host anemones. While this in part may have involved removing mucus and sloughed-off skin, some fishes quite obviously tore off tentacle pieces and ate them. Parts of sea anemones found in clownfish stomachs confirmed these observations.

What benefit does the sea anemone derive from such an arrangement? Because of the strongly developed territorial behavior of virtually all pomacentrid fishes (= clownfishes and damselfishes), these fishes will fend off most other fishes (even those many times their own size) that are about to feed on the tentacles of the sea anemone. In addition, tactile stimulation (discussed above) of the fish by the anemone appears to be a reciprocal event, as the sea anemone also receives similar

In cooler, non-reef environments, perhaps the most familiar example of commensalism is that between a hermit crab and anemones on the shells it inhabits. Several different anemones are involved, many of them species of the genus *Calliactis*, and one species of anemone is not restricted to a particular host hermit crab. The hermits gain protection from the relationship, while the anemones gain food particles left over from the crab's feeding and also a secure substrate that will not become silted over quickly. Photo: K. Giwojna.

Facing page: The clownfish *Amphiprion clarkii* in its anemone, *Heteractis magnifica*. This is a familiar combination throughout the Indo-Pacific. Photo: Dr. H. R. Axelrod.

stimulation from the fish bathing among the tentacles. While direct evidence for this is still sketchy, sea anemones will respond to certain touch stimuli. In fact, a contracted *Calliactis* has been observed to expand again when gently tapped with a finger.

A sea anemone that lives symbiotically with clownfishes derives further benefit from this association, as parasitic crustaceans, dead skin tissue, left-over food, and debris washed onto the oral disc by wave action are removed by the fish.

Underwater film footage has shown repeatedly how certain *Amphiprion* have carried large pieces of food onto the tentacles of their host anemone. It has been suggested that the brightly colored clownfishes act as decoys to attract large fishes that when pursuing the clownfishes become entrapped among the tentacles of the sea anemone. While concrete evidence to sustain this theory is lacking, the implication in terms of mutual benefit for fish and sea anemone would be very significant.

Most marine aquarists will attempt to maintain and observe such fascinating symbiotic relationships in their tank. It is indeed quite easy, provided sound aquarium management procedures are followed. Most readily available are some of the large tropical sea anemones that are the normal home for various clownfishes. In the wild, clownfishes exhibit distinct preferences for particular sea anemones (species and sizes), yet in captivity they also will interact with many other anemones, except burrowing anemones (*Cerianthus* sp.). Sometimes a clownfish appears to be quite desperate, so it is common to find a fish attempting to bathe in the tentacle crown of an anemone with an oral disc that may be smaller in diameter than the maximum length of

the fish. Even temperate sea anemones *(Anthopleura, Tealia)* sometimes are accepted by *Amphiprion* as symbionts.

The clownfish *Amphiprion melanopus* photographed in the anemone *Entacmaea quadricolor* on the Great Barrier Reef of Australia by W. Deas.

A virtually transparent shrimp (probably *Periclimenes*) safe in the tentacles of a Caribbean sunflower anemone, *Stichodactyla helianthus*. Giant anemones often have many commensals in addition to fishes. Photo: P. Colin.

Sea Anemones and Clownfishes

Many of the symbiotic relationships between sea anemones and clownfishes as well as shrimp and hermit crabs, discussed in the previous chapter, can be observed in the home aquarium. Most of the symbionts mentioned are commonly available from marine fish importers and some of the larger aquarium shops. Yet there is one recurring problem, that of proper scientific

This giant anemone is *Heteractis crispa*, recognized by the long, slender tentacles that seem to wave in the current. The clownfish is *Amphiprion chrysopterus*. Photo: Dr. G. R. Allen.

names for the sea anemones. Depending on what literature references are used, the same species of sea anemone may be traded under two or three different names. These name disparities are most glaring between the English and German aquarium literatures.

This problem is further compounded by the already mentioned fact that the entire sea anemone classification needs to be reviewed and overhauled on a global scale. There have been excellent regional attempts made to bring a degree of orderliness into the classification of sea anemones and related forms from particular geographical regions, but large gaps in our taxonomic knowledge about sea anemones still exist.

Fortunately for marine aquarists, there has been

a major effort in recent years to review and clarify the taxonomy of the large tropical sea anemones most commonly encountered by the majority of marine aquarists. The following is based largely on the paper "The clownfish sea anemones," by Daphne Dunn *(Trans. Amer. Philos. Soc.,* 71(1): 1-115), as reviewed in the article "Identifying the giant anemones," by Jerry Walls *(Tropical Fish Hobbyist,* Sept., 1989).

For the benefit of marine aquarists this chapter will provide a brief synopsis of these symbiont sea anemones, as well as the fish and crustacean species often found to be associated with particular sea anemones.

Only a few tentacles of this *Entacmaea quadricolor*, the bulbous giant anemone, display the swollen appearance typical of the species. Although the bulb often is encircled with a band of glandular white tissue (assumedly containing concentrations of nematocysts), this often is not visible. The clownfish is *Amphiprion akindynos*. Photo: W. Deas.

The clownfish anemones belong to several vastly different families. The largest group of symbiont sea anemones—at least eight species—belong to the family Stichodactylidae. The family Actiniidae has only two species of anemones that display a symbiotic behavior with clownfishes, and there is only one known symbiotic species of anemone in the family Thalassianthidae.

The latter is *Cryptodendrum adhaesivum,* a large (30 to 40 cm, 12-16 inches, diameter) sea anemone commonly found in the Indian Ocean, where it lives in association with only a single species of clownfish, *Amphiprion clarkii*. This sea anemone often is the host for shrimp species of the genera *Periclimenes* and *Thor*. It is a very conspicuous sea anemone that tends to cling to the bottom. The oral disc has a thick margin and is covered with a mass of tiny multi-branched tentacles. It inhabits the shallow water region down to a depth of about 3 m (10 feet).

One of the most widely distributed sea anemones throughout the Indo-Pacific region is the bulbous giant anemone, *Entacmaea quadricolor* (family Actiniidae). The aquarium trade has imported this sea anemone under all sorts of names from various parts of the Indo-Pacific. Some of the best-known synonyms are *Actinia helianthus, A. ehrenbergi, Radianthus gelam,* etc. It can be identified easily by a blister-like enlargement (with a whitish colored band around the middle) just before the tip of each tentacle (or at least most of the tentacles). If the blisters are absent, the long tentacles (8 to 10 cm, 3-4 inches, long) will often display a reddish tip.

This sea anemone has been observed from about the Central Pacific (Samoa) to the Red Sea, where it occurs from the shallow inshore waters (near low-tide surface level) down to

about 40 m (130 feet). The maximum diameter of the oral disc of an adult specimen is about 40 cm (16 inches). This anemone is known to be the host for numerous *Amphiprion* species, as well as for the three-spot damselfish, *Dascyllus trimaculatus*.

As mentioned above, there are only two species of actiniid anemone that commonly associate with fishes. The second one is the striped giant anemone, *Macrodactyla doreensis*, which has a fairly restricted range southward from Japan to Taiwan and the Philippines and on to northeastern Australia. It is characterized by a pattern of pale stripes radiating from the oral disc over the tentacles, which are slender and long (5 to 10 cm). The tentacles can be cork-screwed or

The ubiquitous *Amphiprion clarkii* is quite variable in coloration and not at all choosy about what anemone it associates with, especially in the aquarium. Photo: B. Kahl.

wavy in many specimens. The oral disc is relatively small (5 to 10 cm). This sea anemone species is the host for several *Amphiprion,* and possibly also for *Periclimenes* shrimp.

The largest number of clownfish anemones belong in the family Stichodactylidae, the "true" giant anemones. The current literature recognizes two genera: *Stichodactyla,* characterized by very short tentacles, and *Heteractis,* which has fairly long tentacles.

First, let us take a look at those species with short tentacles, the genus *Stichodactyla.* The smallest is *S. tapetum,* which is only 3 to 4 cm (1.2-1.6 inches) across. It is not surprising to find that it is not a host for any symbiotic clownfish. *S. helianthus,* which is common throughout the

This damselfish or humbug, the dominofish *Dascyllus trimaculatus,* has allied itself with a colony of hard corals, *Goniopora,* in an aquarium. Photo: H. Esterbauer.

Caribbean, extending from southern Florida in the north to at least Panama, is the only stichodactylid in the Caribbean. The tentacles of *S. helianthus* are quite short (rarely more than 10 mm, 0.4 inch, long) and form a dense cover over the disc. They are blunt to slightly bulbous at the tip, with a yellowish to greenish color. The margin of the disc often is convoluted. The disc diameter often is about 10 to 12 cm (4-5 inches), but may reach 25 cm (10 inches) in large specimens. There are no clownfish in the Caribbean, but *S. helianthus* plays host for symbiotic shrimp of the genera *Periclimenes* and *Thor*.

The remainder of the stichodactylid anemones accommodating clownfishes occur throughout the Indo-Pacific region. One of the less frequently imported sea anemones is the sticky giant anemone, *S. gigantea*

Because of the length of its tentacles and the often characteristic elbow present, *Heteractis crispa* may be the most readily identified giant sea anemone. Photo: Dr. G. R. Allen.

237

When giant anemones are imported for the hobby, they often look very different from their appearance on the reef. The short tentacles and deeply folded margins of the disc make it likely that this is a species of *Stichodactyla*. Photo: U. E. Friese.

(syn. *S. kenti*). The outer tentacles are about 2 cm (0.8 inch) long, while the inner ones are up to 5 cm (2 inches) long, with roundish or "flame-like" tips. Their color ranges from yellow to greenish, rarely with brighter colors. The oral disc (about 15 to 20 cm, 6-8 inches, in diameter) is strongly folded along the outer margin (contrary to the scientific name, this species is not a giant). A particular characteristic is the constantly vibrating movements of all tentacles. The adhesive action of the stinging cells is so strong (= "sticky"), that tentacles touching and attaching to a human hand, for instance, usually will tear off when the hand is withdrawn.

This species occurs from the Red Sea to New Caledonia and Fiji. In the wild, *S. gigantea* is home for at least six species of

Amphiprion, for *Dascyllus trimaculatus,* and for various crustaceans.

Very similar in appearance to the sticky giant anemone (and often mistaken for it) is the cobblestone giant anemone, *S. mertensii.* This is the largest of all sea anemones, reaching an oral disc diameter of up to 1.5 m (5 feet) and an average size of about 40 to 60 cm (16-24 inches). With such an enormous size, this species is very conspicuous in the wild. The disc is strongly convoluted. The tentacles are very similar to those of *S. gigantea* and are shortest along the outside of the disc (1 to 2 cm, 0.4-0.8 inch) and longest toward its center (up to 5 cm or 2 inches). The nematocysts are not very powerful to the human touch (no "sticky" feeling).

This dominofish, *Dascyllus trimaculatus*, is being cleaned by a wrasse while it hovers with fins extended over the host anemone. Photo: Dr. G. R. Allen.

Large specimens of giant anemones often attract several species of fishes. Here a large *Heteractis magnifica* is hosting several *Amphiprion nigripes*, a *Dascyllus trimaculatus*, and an *Amphiprion clarkii*. Photo: Dr. H. R. Axelrod.

In the aquarium, giant anemones such as this *Heteractis crispa* often are represented only by quite small specimens. Clownfishes may try to occupy even a very small anemone, one smaller in disc diameter than the fish's body length. Photo: C. W. Emmens.

In many areas *S. mertensii* can be recognized on the basis of the *Amphiprion* (clownfish) species inhabiting it. For instance, *A. clarkii*, *A. tricinctus*, and *A. chrysopterus* will take on a darker coloration when inhabiting *S. mertensii*, while they remain lighter in other sea anemones; i.e., the yellow in these clownfish species will be over-shadowed by black when living with *S. mertensii*. If these same fish are then transferred to another anemone they will resume their lighter color. *S. mertensii* occurs from just below the low-water mark down to a depth of about 2 m. Its pedal disc tends to be attached deep in crevices and rocky holes, so that usually only the spectacular tentacular crown is visible.

There is quite a contrast between these two giant sea anemones and some of the other species within the genus *Stichodactyla*. The patchy giant anemone, *S. haddoni* (syn. *Stoichactis kenti)*, for instance, is conspicuous because of its extremely short (3 to 8 mm, 0.12-0.35 inch, long), blunt or rather bulbous tentacles. They are located in radial groups (= patches) on the purplish brown oral disc that can be up to 50 cm (20 in) in diameter. There can be pale and dark groups of tentacles on the same specimen, often with spokes of different colors radiating from the center.

The nematocysts are very powerful; the tentacles stick to human

What determines the color of an anemone is uncertain. There are many factors involved, such as zooxanthellae and other algae, temperature, age, currents, food, substrate, and stress. This *Heteractis crispa* is in a very dull grayish color phase, which could change within hours or days to brighter colors. Photo: W. Doak.

Facing page: Although most giant anemones hide their columns deep within crevices in the reef or in the substrate, *Heteractis magnifica* often lives with the column at least partially exposed. The relatively short tentacles with rounded yellowish tips are fairly diagnostic for the species. Photo: Dr. H. R. Axelrod.

skin (causing welts) and can be torn off. *S. haddoni* occurs over the entire Indo-Western Pacific. Apart from many *Amphiprion* species, it also plays host to *Dascyllus trimaculatus*, some porcelain crabs such as *Petrolisthes ohshimai*, and at least one shrimp *(Periclimenes)*. In some of the literature this species is erroneously referred to as *Discosoma gigantea*.

The second genus of large tropical sea anemones within the family Stichodactylidae is *Heteractis*, which, unlike the previously discussed *Stichodactyla*, has relatively long tentacles.

The largest species within this genus (up to 1 m, 3 ft, in diameter), and possibly also the most attractive of all clownfish anemones, is the fingered giant anemone, *Heteractis magnifica*. This species used to be known in the marine aquarium hobby as *Radianthus ritteri* and *R. paumotensis*. Its broad, usually violet-colored body column is quite conspicuous, even in the wild, since it is commonly found fairly high toward the reef top or platform, where it often can be seen clearly from above the surface. The tentacles are fairly blunt, with bright yellow tips. *H. magnifica* occurs throughout the Indo-Pacific region, where it acts as the host for numerous clownfish species as well as for *Dascyllus trimaculatus* and for several shrimp. Yet *H. magnifica* specimens in the central Indian Ocean (Maldives, Sri Lanka) are occupied only by *Amphiprion clarkii*.

Smaller and less conspicuous is the wavy giant anemone, *H. crispa* (syn. *Radianthus malu, R. paumotensis)*. Adult specimens have an oral disc diameter of about 40 to 50 cm (16-20 inches), usually smaller. Usually only the long (15 cm, 6 inches), pointed tentacles are visible protruding from cracks and crevices, the favorite habitat of this species. In the wild the tentacles of this anemone

Not everything that looks like an anemone is an anemone. Mushroom corals, such as this *Heliofungia* sp., often are deceptively anemone-like, and photos can fool experts. Notice the alternating lines of color on the disc and the sharp edge. Additionally, the tentacle shape doesn't quite fit any common giant anemone. Photo: Dr. H. R. Axelrod.

appear to be flexing with the current (= "wavy"). The coloration of *H. crispa* can be highly variable, from whitish to violet or greenish. The tips of the tentacles are bluish to dark pinkish. It occurs over roughly the same range as *H. magnifica* and acts as host for several *Amphiprion* species, *Dascyllus trimaculatus,* and some shrimp.

A giant sea anemone even more variable in appearance than the one just described is the ringed giant anemone, *H. aurora* (syn. *Radianthus simplex*). Its external characteristics include white spots on the extended, pointed tentacles. These spots can form reticulated patterns or simply series of dots. The white patches are slightly raised, so the large white areas form distinct rounded elevations. Should these extend around the tentacles they produce an alternating pattern of thin and thick patches.

Facing page: An *Amphiprion perideraion* tends its nest in the shadow of a giant *Heteractis magnifica.* Notice the thick, fleshy edge of the disc. Photo: Dr. G. R. Allen.

The usually buried pedal disc and column are orange-red; the column can be totally retracted into the sand. This species is commonly found in shallow, protected water (reef lagoons), and its natural range seems to extend from the Red Sea through the Indian Ocean to the western tropical Pacific. It lives symbiotically with several *Amphiprion* species as well as with *Dascyllus trimaculatus.*

Somewhat less frequently available in the aquarium trade is *H. malu,* primarily because of its small size (7 cm, under 3 inches) and a lack of collectors and/or suppliers. This species seems to be concentrated in the tropical Pacific. Its tentacles are—by comparison with the previous species—fairly short (up to 4 cm, 1.6 inches) and often uneven. It appears to associate only with *Amphiprion clarkii* and *Dascyllus albisella.*

These are the principal clownfish anemones. As mentioned before, clownfishes will accept substitute sea anemones. In fact, in my many years as a marine aquarist I have seen *Amphiprion* accept just about any other tropical or even temperate sea anemone that was kept in the same tank with them. The only sea anemones that appear to be largely ignored by *Amphiprion* are those members of the family Actinodiscidae that have extremely short (1 mm) or virtually no tentacles, primarily the small *Actinodiscus* species. Some of the larger *Rhodactis* species seem to be acceptable to *Amphiprion* as hosts, certainly in captivity. A word of caution, though: *Rhodactis* is a known predator of other reef fishes, including surprisingly large specimens that have died when coming into contact with the nematocysts of these anemones.

As a final word about clownfish anemones I

A group of dominofish, *Dascyllus trimaculatus*, finds refuge and a home in a giant *Heteractis*. Photo: W. Deas.

must once again state that it often is very difficult, sometimes even impossible, to correctly identify some of these large tropical sea anemones on the basis of coloration of the oral disc and the tentacles. One reason for this is the presence of vast numbers of unicellular algae (zooxanthellae) within the tissue of the anemones. The density and colors of the algae are determining factors for the various color shades even within different animals of the same species.

On the other hand, the column wall and the pedal disc are hardly affected by these algae, so these together with the species-specific warts (if present) provide far better characteristics for proper identification. Since many of these sea anemone species have most of their column buried in coral rubble, sand, or crevices, often the only identifiable features are the oral disc and the tentacles. Fortunately, their shape,

length, and size often are species-specific or at least genus-specific, which then makes a proper identification quite possible from the characteristics listed in this chapter.

Facing page: These *Amphiprion tricinctus* are hovering over a rather small *Heteractis aurora*, one of the less common giant anemones (only 15 cm in diameter). Notice the constrictions on the tentacles where the white rings bulge out a bit. Except for not having an obvious central mouth or brighter colors, there is a lot of similarity to *Phymanthus* species. Photo: Dr. G. R. Allen.

Sea Anemones in the Miniature Reef Aquarium

Keeping sea anemones in home aquaria has been done more or less successfully for many decades. In the past there was the typical marine tank set up according to principles and guidelines laid down in many good "how to" marine aquarium handbooks. These texts elaborated in great detail on the composition of suitable sea water (natural

Facing page: With care and the proper equipment, success with anemones and other cnidarians can be yours. Photo: P. Wilkens.

and synthetic) and the principles, techniques, and equipment used to maintain the quality of sea water with marine fishes and invertebrates present in the tank.

As time went by the knowledge of precisely what happens to sea water in closed systems—as it is exposed to the metabolic waste products of the aquarium inhabitants—improved considerably. At the same time (most notably during the last 20 years) water treatment technology and equipment improved dramatically. This process has accelerated even faster during the last ten years or so.

These unprecedented technological advances in the marine aquarium hobby were, in part, due to the obvious need to provide better and more skillful care for marine invertebrates. These had been coming onto the aquarium market in ever-increasing numbers and greater species diversity, and they quickly proved to be far more sensitive than even many of the most delicate coral reef fishes.

The crowning achievement for most marine aquarists is to keep live coral successfully and for long periods of time. Many tried, but most failed after a relatively short period of time. Marine aquarists began to search for more natural ways to keep these and other delicate marine invertebrates. Clearly, aquarium conditions had to come closer to those prevailing in the wild, where there was essentially a dynamic equilibrium, a balance of all those factors that essentially sustain life in the sea or—more appropriately for marine aquarists—on a coral reef.

The major factors involved, such as maintaining water quality by recycling nutrient materials through appropriate trophic levels, sufficient light as the best energy source, and an animal population that would not jeopardize the

delicate balance between them and their environment, needed to be not merely simulated but actually duplicated in the marine aquarium. Early experimental work by marine aquarists in Europe, most notably Germany and Holland, which was facilitated by advanced water treatment and new aquarium technology, led to the concept of the miniature reef aquarium (NOT to be mistaken for the "Minireef" trademark). Since then this concept has proven to be the only realistic way to keep not only coral, but indeed all marine aquarium animals, and most certainly sea anemones.

In order to explain what a miniature aquarium is and how it supports the well-being of all animals kept in marine aquariums, we need to look briefly into the historical development of this unique marine aquarium concept.

Many marine aquariums actually are rather dull, with grayish non-descript invertebrates and a few fishes. This tank, though obviously successful, could do with a bit more color, don't you think? Photo: H. Esterbauer.

257

THE MINIATURE REEF AQUARIUM—AN ABRIDGED HISTORY

As mentioned above, the miniature reef concept evolved from attempts to keep live coral in closed-system aquaria (such as the home aquarium). The first success was reported by an Indonesian aquarist, Mr. Lee Chin Eng, in 1961. He recommended the use of untreated (raw, but clean) sea water in a tank filled with naturally overgrown rocks taken from a reef. To this he would then add algae and a range of invertebrate animals including live coral. This early attempt does not refer explicitly to light as an essential ingredient, but its usefulness appeared to have been tacitly implied. Here we find one of the early references to the "natural system."

During the following years there were a number of articles on how to best keep coral under "natural" conditions. This was followed by considerable controversy among some of the authors, but there also were notable aquaristic successes with live coral, as well as with other marine invertebrates. The first explicit reference to the use of intense light as part of this natural system did not come until 1971. This point was further emphasized by a description of a system in which live coral was exposed to direct sunlight for a few hours each day, so that the ". . . symbiotic zooxanthellae manufactured sufficient food to sustain their (coral) host."

While this development explained the food and feeding of live coral under aquarium conditions, it did not address the removal of waste products, which must occur in a closed-system home aquarium. In the late 1960's we find the first detailed references to algae kept under intense illumination for the removal of nutrient material from aquarium water. In fact, in 1968 we find one of the first

detailed descriptions of algal filters and the periodic cropping that is required as the algae grow, utilizing the nitrogenous compounds (principally nitrate) in the water.

Work on algal filters and the utilization of algae in aquarium systems continued into the early 1970's, especially in conjunction with the natural system of keeping live corals in aquaria.

Some of the early references to these systems as being "mini-reefs" appeared in the aquarium literature somewhere around the mid-1970's. In essence, at that time all miniature reef aquariums were characterized by the use of large masses of living rock and strong algal growth. Incidentally, the use of "living rock" as a critically important element in the

In a well-balanced marine aquarium it is possible to keep sessile invertebrates, active invertebrates, and a few colorful fishes. This colorful porcelain crab, *Petrolisthes maculatus*, is in a *Stichodactyla haddoni*. Photo: U. E. Friese.

The advent of modern high-intensity lighting of several types, including actinic and mercury vapor lights, has made possible the keeping of many invertebrates, anemones included, that contain symbiotic algae. Photo courtesy Energy Savers.

natural system (or by then, the miniature reef) continued throughout this period.

To this day it is not known who actually coined the term "living rock," but it has become firmly established among marine aquarists, especially those wanting to keep delicate marine invertebrates such as sea anemones. At this stage it may be appropriate to correctly define the term living rock; the most appropriate definition seems to come from Bruce A. Carlson of the Waikiki Aquarium. He proposes the following:

"Living Rock . . . a piece of carbonate reef substrate formed by the action of living organisms such as corals and coralline algae and con-solidated as part of the reef framework after the death of the original coral/coralline algae and continuing to provide shelter and a growing surface for numerous reef plants, invertebrates, fishes, and subsequent generations of corals and coralline algae."

A properly functioning miniature reef aquarium, especially one that is to accommodate sea anemones, should have a solid base of living rock.

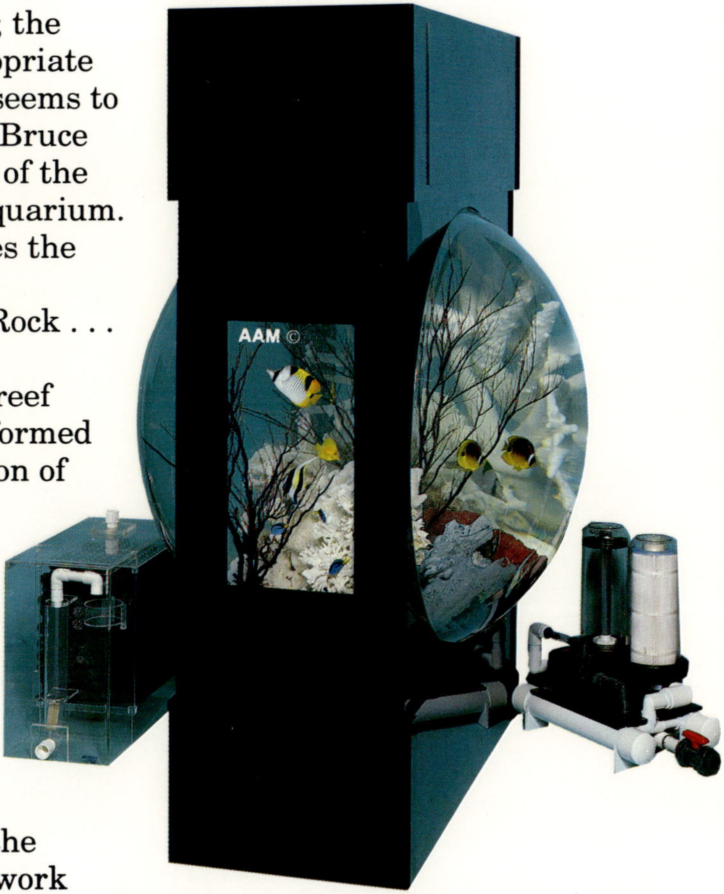

A fantastic complete marine aquarium system suitable for a small museum. Today acrylics can be made into sturdy marine tanks of many special shapes. Photo courtesy American Acrylic.

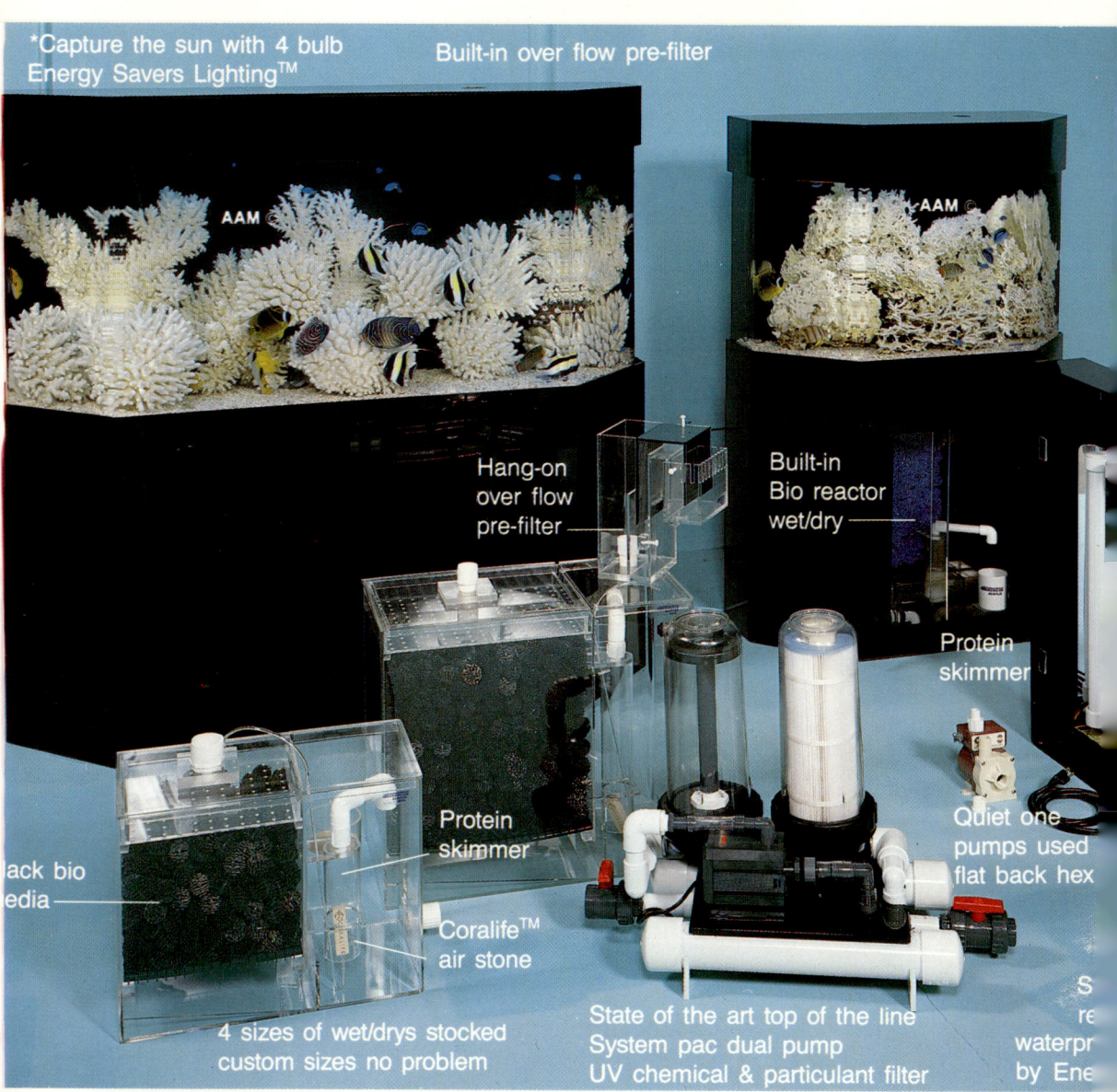

HOW TO SET UP A MINIATURE REEF AQUARIUM FOR SEA ANEMONES

For obtaining basic instructions for setting up a marine tank, the aquarist is best advised to follow the details laid out in any of a number of "how to" aquarium handbooks, such as Emmens's *Marine Aquaria & Miniature Reefs* (TFH). In order to turn the basic marine aquarium into a functioning miniature reef tank, the following five basic elements have to be included:

1) Sea water (natural or artificial) of highest quality.

2) A trickle filter, which must have a volume of at least 5% to 10% of the total aquarium volume.

3) Strong illumination, particularly from that part of the light spectrum essential for strong algal growth.

4) The correct temperature range and proper temperature control.

5) Substantial growth of higher algae and a low population density of invertebrate animals (none of these can be herbivorous).

No matter how effectively the above five factors are made to work together, an aquarium is still an artificial environment, so certain supportive measures should be taken to assure optimum conditions. These measures should include:

6) Strong water currents (= optimum water circulation).

7) If possible, alternating flow direction (i.e., essentially circular currents; six hours to the left and six hours to the right, to simulate tidal flow).

European aquarists often go yet another couple of steps further, especially when corals are kept. These steps include carbon dioxide diffusion in order to enhance CO_2 uptake by the zooxanthellae inside coral polyps (and in sea anemones), maintaining

Facing page: A display of some of the most up-to-date equipment for the miniature reef aquarium. Today's tanks and equipment are streamlined, sturdy, and efficient. Photo courtesy American Acrylic.

proper iron concentration for the metabolism of bacteria (working in the trickle filter), and minimum water changes (not more than 10% per month).

The following equipment should NOT be used in conjunction with a miniature reef aquarium: no protein skimmer; no air stone; no ozonizer; no UV lamp. MOREOVER, it is absolutely imperative that the aquarium population of fishes and/or invertebrates be specifically selected for species and specimens that do not have an adverse effect on the miniature reef habitat. They must not graze on the algae, pick on corals, sea anemones, and other invertebrates, AND must not overburden the entire system with their metabolic waste products. Only a few small fishes and suitable invertebrates are permitted in a fairly large tank.

Now let us look at the principal factors in detail to see just how they transform the tank into a miniature reef aquarium for sea anemones.

SEA WATER

Sea anemones, like all other marine invertebrates, are sensitive to changes in their environment. Removing these animals from the sea, with its relatively constant water quality, and transplanting them into a home aquarium is a very drastic environmental change for them. In such a move many things have changed for sea anemones, but the sea water (or rather the quality of the sea water) is the singular most important factor.

Sea water is a mixture of salts in nearly constant proportions that, depending upon the location in the sea, is combined with a variable amount of fresh water. This then determines the overall salinity in sea water. Although most known chemical elements of the earth are found in sea water, many of them occur only as trace elements, but most of these are vitally

important to the survival of most animals. The bulk is made up of six elements that comprise more than 99% of the sea salts. These are chlorine, calcium, potassium, magnesium, sodium, and sulfur. Since water in the oceans is well mixed, the relative abundance of these major components of sea water is essentially constant.

This concept of constancy of composition is of great importance in the marine environment. It is the very basis for the fundamental stability to which all marine organisms have adapted. To the marine aquarist this means that his animals, especially the invertebrates such as sea anemones and related

With almost no exceptions, sea water from around the world contains the same elements and compounds in the same proportions. A good artificial salt mixture must be equally uniform. Photo courtesy Tropic Marin.

forms, have very little, if any, tolerance for drastic environmental changes. So, the sea water provided for sea anemones in captivity has to be of excellent quality.

For those marine aquarists not living by the sea, obtaining high quality natural sea water is a major problem. Even those who are lucky enough to live in close proximity to the sea may have problems getting good sea water. Inshore waters often are heavily diluted with freshwater (rain) run-off or, most likely, the sea water is polluted by industrial and domestic waste products. Clearly, then, for many aquarists the answer is to use artificial sea water made from commercially available salts mixed with distilled or certain mineral-rich spring waters. Most aquarium and pet shops will have a range of name-brand sea salts available. The instructions for mixing the salts usually are quite explicit and easy to follow.

An important point to remember is that artificial seawater must be sufficiently aerated and filtered over charcoal (to remove any impurities) PRIOR to use in aquaria.

Moreover, newly mixed artificial sea water lacks specific biological activity, and accumulated waste products from the tank's inhabitants are not removed by normal filtration. In order to overcome this problem it is advisable to seed the new tank with either water or some filter medium from an already established marine tank. Some books suggest placing a handful of garden soil in the artificially mixed sea water overnight. The next day most of the sand and debris will have settled to the bottom, and the supernatant fluid is then siphoned off, filtered, and used in the aquarium. This is a dangerous practice since most soil around human habitation usually is contaminated by undesirable, even dangerous, chemicals.

Trickle filtration systems are among the most dependable methods for providing filtration in a marine aquarium and are especially beneficial in tanks housing sea anemones. Photo courtesy of Aquarium Life Support Systems.

Instead, it is advisable to seed with water from an already established tank.

TRICKLE FILTER

One vital element in a successful miniature reef aquarium, whether it is used for sea anemones or other invertebrates or even for fishes, is an effective trickle filter. Trickle filter technology, heralded in some of the recent aquarium literature as something new and revolutionary, is actually quite old. It has been (and still is) used on a large scale in commercial sewage treatment plants. How could such an old and proven effective biological filtration method have gone unnoticed by aquarists for so long? In order to provide the answer to this question we have to take a brief look at the processes involved in biological filtration.

Diagrammatic views of a trickle filter, with insets of the biologically active medium, in this case fine coral rubble. When used correctly, trickle filters provide water quality that is adequate for keeping many delicate marine invertebrates. Art: J. R. Quinn.

The word filtration implies by common consensus that this process involves the mechanical, selective retention/removal of solid particles from some other substance that is being filtered. In our case it suggests that certain materials are being mechanically removed from the water in our fish tank. This is not quite the way it works. A more appropriate term for the concept of biological filtration would be "biological degradation and assimilation."

We know that bacteria generally are responsible for the decay of dead organic material, be this on land or in water. In marine and freshwater aquaria the organic materials we are concerned with as aquarists are primarily the nitrogenous waste products given off by the aquarium's inhabitants. The bacterial degradation most beneficial to an aquarium—especially to a miniature reef aquarium—is the sequential conversion of complex organic, nitrogenous waste materials to nitrate, an inorganic compound that is basically a fertilizer.

This process must be encouraged if an effective and fully functional miniature reef aquarium is to be established. It assures that various more or less toxic substances are transformed into the relatively harmless nitrate, which acts as a nutrient for the higher algae and for water plants in general. This process is accomplished by bacteria of the genera *Nitrobacter* and *Nitrosomonas*. Its efficiency depends very much on the living conditions of these bacteria. They are aerobic; that is, they must have oxygen to live and reproduce, so they must be provided with an environment that has an abundant oxygen supply.

Bacteria are either motile (move about constantly) or sessile (living attached to some solid object). *Nitrobacter* and *Nitrosomonas* are sessile bacteria that can live attached anywhere in the aquarium; few are ever motile (moving about freely in the water) at any particular time. The largest numbers are in those locations where there are the most favorable living conditions, where there is an abundant food and oxygen supply. This is clearly where there is the highest flow rate of water in an aquarium—inside a filter with its constant flow-through of nutrient-rich water.

There is a vast range of commercially available filters on the market, from subgravel filters to a dazzling array of power filters. Yet conventional filter media (either the tank's gravel bottom or the substance such as gravel or charcoal used inside the pressure vessels of power filters) do not provide a large enough surface area for sufficiently large bacterial populations to become established. Moreover, the filter medium has to be sufficiently coarse that it does not become plugged up by dirt particles and debris from the aquarium water.

Another factor in an efficient filter operation is an abundant and uninterrupted oxygen supply. Normally, when the water stops flowing through a filter (subgravel or power filter) the oxygen supply is interrupted, and after a surprisingly short period of time the "useful" bacteria begin to die off and less favorable ones develop (i.e., those that can live without oxygen), which can quickly kill off all tank inhabitants.

How do we assure that the oxygen supply does not stop and that the filter medium does not become caked over with debris? In other words, how do we guarantee optimum conditions for the denitrifying bacteria in our miniature reef aquarium filter? We use a trickle filter.

There are a number of different versions and models of trickle filters, with the various

Although the sophisticated marine aquarium system may be technologically complex, it is not necessarily ugly. Many manufacturers can provide self-contained units that will fit the decor of any room. Photo courtesy American Acrylic.

manufacturers praising the particular advantages of their own models. The two essential functional elements of a trickle filter are a very coarse medium (one that provides maximum surface areas for bacteria to become established on) and a surface area that is completely exposed to the air but still permits an uninterrupted gravity-flow of water.

While there are many different models of trickle filters, there really are only two different methods of operating them. In the first situation, the water flows through an overflow from the tank to enter the filter. It then passes through the filter under gravity and is then collected in a reservoir at the bottom of the filter, from where it is returned to the tank by means of a circulating pump. The second method places the trickle filter above the tank. The water is pumped from the aquarium to the trickle filter with a pump and after passing through the filter returns to the tank by gravity flow. The actual plumbing connections can be in a number of configurations to take best advantage of particular tank sizes and locations. The trickle filter even can be remote from the tank it services.

I favor the second method, simply because a trickle filter easily can be located above a tank (built-in neatly in a cabinet that provides sufficient overhead room). Moreover, the returning water can be discharged directly into the tank below through one or (preferably) more down-pipes, thus providing additional aeration to the tank. A spray-bar (a perforated pipe) for the water coming from the tank will, when fitted over the entire length of the trickle filter surface, further enhance the filter's performance. Some authors suggest the additional use of a separate fine filter

Healthy anemones, such as these tube anemones, *Cerianthus*, provide attractive backgrounds for more active invertebrates such as the starfish *Patiria miniata*. Be aware that starfishes often are serious predators on anemones. Photo: Dr. H. R. Axelrod.

(cartridge filter) to prevent undecomposed debris and particles from returning to the tank. The same result also can be achieved easily and inexpensively by placing a foam rubber mat over the surface area of the trickle filter, which will retain larger particles. This mat can be quickly rinsed out for renewed use.

There has been, and still is, much debate about the most effective medium for trickle filters. Irrespective of any manufacturer's claims, a number of inert substances can be used. Remember that the objective is to provide maximum surface area for bacteria to become attached to and to make

The miniature reef tank today is high in lower invertebrates, especially anemones and other cnidarians, and has only a sparse representation of fishes. Photo: J. Burleson.

Facing page: Left: Purified carbon is highly adsorbtive and removes impurities from the aquarium water. Photo courtesy Rolf C. Hagen Corp. **Right:** Biospheres and other filter media devised for use in trickle filtration systems greatly expand the amount of surface area available as living space to which the bacteria that the system depends upon can attach themselves. Photo courtesy Rolf C. Hagen Corp.

an adequate oxygen supply available to these bacteria.

While everything from small river stones to marble chips and even glass marbles has been used quite effectively in trickle filters, one of the best media is "bio-balls" or "plastic hedgehogs." Basically, these are spherical filter scaffoldings or hollow matrices made of polyethylene. A lot of quantitative and qualitative research has been done on "bio-balls" as a medium in trickle filters. A typical "bio-ball" is 40 mm (1.6 inches) in diameter and has a complex internal lattice of plastic that provides an enormous surface area. A trickle filter of 20 liters (5.2 U.S. gallons) volume (which could serve a 400-liter or 104-gallon aquarium), filled with "bio-balls" provides a surface area of about 80,000 square centimeters (12,800 sq. inches!). That these plastic spheres are essentially hollow is another significant advantage, as it enables an even flow of the aquarium water (no channelling effect), and air can freely circulate through the filter medium to provide the oxygen used by the bacteria for decomposing the nitrogenous waste material.

Other materials also can be used in trickle filters. For miniature reef aquaria I have used very coarse coral rubble and even medium-coarse marble chips. In addition to satisfactory filtering properties, in a marine tank such carbonate-rich media also provide a stabilizing effect for the pH value. It has to be admitted that "bio-balls" assure, no doubt, one of the best overall flow-through rates while also providing a maximum surface area on which bacteria can become established.

For proper biological filtration it is generally recommended to use a trickle filter that is 5% to 10% the volume of the tank to be filtered. The volumetric dimensions can

Facing page: High-tech lighting, in this case actinics, provides the quality and intensity of light necessary for the survival and even reproduction of many anemones and other cnidarians. Photo: J. Burleson.

be lengthwise or vertical, whatever space configuration is available and how it can best be disguised. As mentioned before, this author has used overhead trickle filters made to order out of sheets of glass glued together and built into the lid of the display unit. Water is pumped from the bottom of the tank either through the substrate (coral rubble) or from a position above the substrate, and the circulating pump also usually is housed in the bottom of the cabinet. Conventional plastic (PVC) pipes supply the overhead trickle filter. When designing the layout for the plumbing it is absolutely essential to keep in mind the disastrous consequence of a power failure if the water flow does not stop automatically!

There has been much debate in the aquarium literature about the use, effectiveness, and best construction for trickle filters, most notably about the required size (volume), the medium, and the general efficiency. While trickle filters no doubt are one of the single most important factors in establishing a functional miniature reef aquarium, it can be argued that these filters can be too efficient and significantly distort the very "balance" they are designed to establish and maintain. Only very close monitoring of the water conditions and appearance of the miniature reef aquarium can ultimately assure success. Nevertheless, sea anemones have constantly shown that they thrive best in a marine tank equipped with an effective trickle filter.

LIGHTING FOR THE MINIATURE REEF AQUARIUM

Strong illumination is absolutely essential for a successful miniature reef aquarium, especially one that is intended for sea anemones. Why is this so, and what kind of illumination are we talking about here? In order to provide the answers we have to take a brief look at what light, especially sunlight, really is and how it has the profound effects on plants that it does.

The single most important natural energy source on earth is sunlight. What are its effects on the aquatic environment? The transfer of energy from the sun to the earth, in our case to the ocean, is known as solar radiation. Some of this radiation is reflected and part is absorbed at the sea surface. It is the portion that is absorbed that concerns us most as aquarists.

The solar radiation occurs over a wide spectrum of electromagnetic waves emitted by the sun, yet that part of the spectrum that is of importance to

This heavily "planted" aquarium contains many pounds of living rock designed to look like a coral reef background. Notice the paucity of fishes in this setup. Certainly it could use more color to accent the rubble background. Photo: Dr. H. R. Axelrod.

the biological processes in the sea is quite narrow. The visible electromagnetic frequency range is called "light." It is measured in nanometers (nm), 1 nm = 0.000,000,001 mm. The range of optical radiation extends from 100 nm to 1,000 nm; humans can see light from about 380 to 780 nm. Various nanometer ranges correspond to particular colors of the spectrum: 380 nm to 440 nm = violet; 450 to 490 nm = blue; 500 nm to 560 nm = green; 570 nm to 590 nm = yellow; 600 nm to 630 nm = orange; 640 nm to 780 nm = red. The invisible radiation that we recognize as heat begins above 780 nm. At the lowest end of the range, below 380 nm, ultraviolet radiation (which is damaging to the human eye) commences.

The electromagnetic wave range of greatest significance to marine aquarium technology extends from about 350 nm to 500 nm. Referring to the information above, we can see that this light is violet to blue. Biological oceanographers have found that this light can easily penetrate the sea surface because scattering of the light waves due to the water molecules is minimal, yet penetration can easily be impeded by colored substances in the water.

This interesting aquarium group includes the tank shown on the previous page as well as a shallow coral and algae component. Notice the use of eight separate light units to power the biological activity in the aquariums. Photo: Dr. H. R. Axelrod.

A corallimorph anemone, possibly one of the *Pseudocorynactis* species. Photo by Courtney Platt.

Light waves from the lowest end of the violet range (around 380 nm) can penetrate to about 100 m in very clear oceanic sea water.

Penetration into water varies among the different light sources. Those from the lower end of the spectrum (shortest waves) have the largest energy, while those from the upper end (longest waves) of the range have the lowest energy. From this it follows that, for instance, yellow and orange light sources (above 550 nm) are strongly impeded when entering water, and they have a tendency to heat up the water. How can we relate all this to sound miniature reef aquarium practices?

The very basis of a miniature reef tank is a luxuriant growth of higher algae. Many sea anemones—just like corals—accommodate intracellular algae (zooxanthellae) within their tissues. In addition, there are various lower and higher algae (e.g., *Caulerpa*) present in a successful miniature reef aquarium. With few exceptions, most of these algae can not utilize long light waves. For marine aquarium purposes (and, incidentally, also for freshwater aquaria) a composite of light sources will have to be provided if we are to establish a functioning miniature reef aquarium for sea anemones.

Unfortunately, the normal light mixture in the natural habitat of most marine organisms lies in the range from 350 nm to 580 nm (with peak efficiency at about 480 nm), so there is not too much to choose from in terms of commercially available light sources that can be used for a miniature reef aquarium. The best results have been achieved from a combination of halogen vapor lamps with special-performance fluorescent tubes.

The products most commonly used by marine aquarists are HQI mercury vapor lamps. These should be supplemented by fluorescent tubes producing light in the violet/blue range. For that

A marine tank that can sustain anemones also should be able to grow a vigorous crop of marine algae (seaweeds). These *Caulerpa* fronds provide color, food for some fishes and invertebrates, and excellent biological activity. Photo: Dr. H. R. Axelrod.

A colorful zooanthid from off the coast of California. Photo by Courtney Platt.

purpose actinic 03 fluorescent tubes have proven to be very effective. These tubes produce an intense but very narrow electromagnetic wave form that reaches it peak at about 420 nm. This sort of light is particularly useful for the zooxanthellae in the tissues of sea anemones and corals. In fact, actinic lighting can also be used in combination with normal daylight fluorescent tubes.

Actinic 03 tubes come in 0.6 m (2 feet) and 1.2 m (4 feet) lengths. How many of each should be placed together with a light source that produces light waves up to about 550 nm is difficult to say and should be a matter for some experimentation for best results. It should be noted here that an excess of actinic lighting will produce a distinct blue lighting that is clearly appreciated by many sea anemones and coral polyps but gives a marine tank a very unnatural appearance. Adding a strong halogen vapor or daylight fluorescent source quickly overcomes this problem. The result invariably is a superb daylight effect!

References to light intensity (lux) and light flow (lumen), often mentioned in some detail in aquarium discussions about lights and lighting, deliberately have been avoided here. The reason for this is simple: there are as yet no suitable metering devices available for use in aquarium operations. Consequently, aquarists have no monitoring capability (except relying on the often scant technical information provided by manufacturers). Those lux meters commercially available are standardized for the spectral range of the human eye (around 550 nm median). Aquarists would require measuring devices for the ranges 315 to 500 nm and again from 600 nm to 700 nm (for use in freshwater tanks). Consequently, much of the data base for lux and lumen available from the aquarium literature is virtually useless.

PREFERRED TEMPERATURE RANGES OF SEA ANEMONES

Unlike the highly variable temperatures in rivers, lakes, streams and ponds, the oceanic environment has relatively stable temperatures. Certainly there are minor seasonal variations—as on the Great Barrier Reef in Australia and in the Red Sea and most other oceanic regions—as well as some inshore temperature fluctuations due to tides, freshwater run-off, etc. Yet, by-and-large, marine invertebrates such as sea anemones are not exposed to dramatic temperature fluctuations.

Closeup of the tentacles in a *Condylactis* species. Photo by Courtney Platt.

Actinic 03 lamps are considered the best all-around type for a general marine aquarium. Their blue light is appreciated by many cnidarians and other invertebrates. Photo courtesy J. P. Burleson, Inc.

Clearly, most marine aquarists will keep tropical species of sea anemones, be these from Hawaii, Micronesia, Taiwan, the Red Sea, Florida, or the Caribbean. Many marine aquarists also will keep temperate marine tanks, where the prevailing temperature ranges can be quite different (i.e., much lower).

(Yes, a miniature reef aquarium does not always have to be a heated tropical tank. Temperate and even cold-water marine tanks can be operated along the same miniature reef principles discussed above.)

Before we take a quick look at some of the regional temperature ranges that marine aquarists may have to provide for their sea anemones, it must be pointed out at this stage that it is vitally important to ascertain from the dealer or collector the origin of a particular anemone. Neglecting this point can lead quickly to the demise of the specimen!

Preferred temperature ranges, for

sea anemones from:

Hawaii: 22 to 27°C (72-81°F); surface waters uniformly warm.

Great Barrier Reef: 22 to 27°C (72-81°F).

Red Sea: 25 to 28°C (77-82°F).

North & Central Indian Ocean: 22 to 25°C (72-77°F).

Micronesia: 22 to 27°C (72-81°F); surface waters uniformly warm.

Gulf of Mexico: 22 to 25°C (72-77°F) (northern Gulf); 23 to 27°C (73-81°F) (southern Gulf).

Gulf of California: about 15°C (59°F) (northern Gulf, avg. winter temp.); about 20°C (68°F) (southern Gulf, avg. winter temp); up to 28°C (82°F) during the summer months.

Caribbean: 25 to 28°C (77-82°F); surface waters uniformly warm.

Southern Florida: 23 to 26°C (73-79°F).

Atlantic coast from Cape Cod to central Florida: 15 to 23°C (59-73°F) (highly variable).

Nova Scotia to Cape Cod: 12 to 15°C (54-59°F).

Alaska to northern Mexico: 8 to 14°C (46-57°F); slightly warmer off southern California.

Puget Sound and adjacent waters: 8 to 12°C (46-54°F).

These are some of the representative temperature ranges for areas from which sea anemones are brought into the aquarium trade or are collected by hobbyists. If in doubt about the origin of a species, it is best to stay on the LOW side as far as temperature is concerned. This is invariably far less harmful than temperatures that are in excess. Many sea anemones can tolerate surprisingly high temperatures, but this places considerable stress on the animal as well as an additional burden on the miniature reef aquarium system (less dissolved oxygen in the water, significantly increased CO_2 uptake by the algae, increased metabolism of the filter bacteria).

MAXIMUM POPULATION LEVELS IN A MINIATURE REEF AQUARIUM

Even a well-functioning miniature reef aquarium is at best in a precariously balanced situation that easily can be upset. We must make sure that whatever we place in it does not disturb this balance. From the beginning we must clearly avoid any tank inhabitants that have the potential to jeopardize the miniature reef system. The essential criteria we must be striving to obtain are intra- and interspecific compatibility and a population level (i.e., number of animals) that is sustainable in our particular miniature reef aquarium.

Although in this book we are principally concerned with sea anemones, most marine aquarists also want to keep other animals together with sea anemones. Most of these probably will be other invertebrates, but they

If you want to keep clownfishes such as these *Amphiprion nigripes*, you will need a tropical reef tank and anemones that are adapted to higher water temperatures. Many anemones are cool- or even cold-water animals. Photo: G. Spies.

may also be mixed with a few fishes (as for instance, large tropical sea anemones together with clownfishes). Our first priority when selecting animals for the miniature reef aquarium is to make sure that they get along among themselves and that they do not interfere with the basic system supporting the miniature reef aquarium. Basically, there must be NO ALGAE EATERS! It should be noted here that many molluscs (e.g., cowries) feed voraciously on algae, which would greatly diminish the nutrient recycling capability in the miniature aquarium.

It also must be remembered that, while many sea anemones have intracellular algae in their tissues, they do not rely exclusively on photosynthetic production of food. Many of the particle feeders and predators must be given supplementary feedings. This adds a further burden on the miniature reef aquarium system.

Fish numbers and sizes must be kept to a minimum. They produce large amounts of nitrogenous metabolic waste materials that can quickly exceed the system's capability to handle the increase effectively. This rapidly can lead to an over-all breakdown of the miniature reef aquarium. What constitutes a "minimum" is a difficult question to answer. It depends principally on the effectiveness of the trickle filter and lighting, as well as on the extent of algal growth in the tank. Clearly the marine aquarist operating a miniature reef aquarium will have to closely monitor the system by checking the pH and dissolved oxygen, carbon dioxide, ammonia, nitrite, and nitrate levels.

Now a few words about COMPATIBILITY of sea anemones and related animals in a miniature reef aquarium. Although the majority of these animals are more or less

sessile (although many of the sea anemones can and will move about quite freely), this does not mean that there is no competition for suitable sites for attachment, which often leads to the demise of the losers.

The fundamental rule is that in a miniature reef aquarium (in fact, in any marine tank) all solitary sea anemones must be placed in such a way that, when fully opened up and expanded, they can never touch each other. When this rule is followed it will—considering the size of some of the giant sea anemones available—quickly provide a natural limitation on the numbers of animals placed into the tank. Small colonial or encrusting sea anemones, such as *Parazoanthus*, often inhibit the spreading of other cnidarians by their mere presence. In particular, the reddish

Most anemones, including this *Heteractis magnifica*, are relatively defenseless against determined predators. Their stings will kill or incapacitate small animals, but many fishes and crabs consider anemones a delicacy at all costs. Photo: U. E. Friese.

Facing page: Many anemones compete for space and light by actively stinging competitors that get too close. Thus, make sure that every anemone in the aquarium is separated from every other anemone by a sufficient distance to prevent "fights." Photo: B. Kahl.

brown mushroom anemones of the genus *Actinodiscus,* when growing toward other sea anemones and corals, are known to gradually damage and kill these. So extra care has to be taken when these popular colonial anemones are placed in a miniature reef aquarium. Small sea anemones, such as *Aiptasia* and the small *Palythoa* species, often exhibit almost explosive growth in a miniature reef aquarium. This can lead to a severe shortage of suitable sites for other sea anemones and severely hamper their development.

Soft corals (e.g., *Xenia* and *Anthelia)* also can take over a miniature reef aquarium and reduce the space available for other animals. Many coral species can and do inflict severe damage to each other should they come into direct contact. Special care has to be taken with these when placing them in a miniature reef aquarium. However, most reef corals (e.g., *Goniopora)* seem to get along quite well with each other, and even contact between the polyps from adjacent colonies seems to do little if any harm.

So far we have discussed the five basic and most important elements needed for an effective and successful miniature reef aquarium that is suitable for sea anemones and other invertebrate and vertebrate animals. There are a few other measures that should also be taken in order to assure continuous, optimum conditions in the tank.

WATER CURRENTS AND CIRCULATION

Physical oceanic conditions are characterized by strong currents, both periodic tidal currents as well as vast water movements on a global scale. In addition, there are a host of local conditions all resulting in some form of water movement. In fact, coral reefs must have certain minimum water currents

flowing over and around them in order to survive. Early experiments with live coral displays demonstrated the effectiveness of water currents; the polyps would expand more readily and stay out longer, and the overall longevity of live coral could be increased quite dramatically. Since then the knowledge of water currents in the aquarium has increased substantially and the supporting technology has reached virtual perfection. In order to maintain optimum water conditions in a miniature reef aquarium, there should be an adequate supplementary circulation system.

Exactly what happens when water is properly circulated through the miniature reef aquarium?

—The circulating water assures an even temperature distribution throughout the entire aquarium. There are no hot or cold pockets.

—Dirt particles, food remnants, etc., are kept in suspension until they are picked up by the filter and then are physically removed and biologically degraded and absorbed.

—Oxygenation reaches and maintains maximum saturation levels.

—Water currents require greater physical activity, most notably by the fishes, enhancing the demand for food and at the same time improving metabolism (through exercise). This also facilitates optimum development of muscles and fins.

—Prepared fish foods are distributed faster and maintained longer in suspension (= better and longer feeding; more cost effective).

—Liquid or very fine food for invertebrates is better distributed throughout the tank and assures better and more reliable feeding by those animals that must wait for the food to come to them. This may be the single most important factor for the new settlement of sessile

By using the currents in the aquarium to your advantage, you can help circulate food, especially the variety of liquid invertebrate foods now available in pet shops, through all parts of the tank, letting small anemones settle wherever they can find a good substrate. Photo courtesy Coralife, Energy Savers.

invertebrates on previously clean rocks and other solid surfaces.

—More constant and uniform distribution of trace elements (e.g., iron) throughout the tank.

The two most important factors in water circulation on a natural coral reef are no doubt the provision of nutrients and oxygen and the removal of metabolic waste products. Also of great importance is the physical stimulus provided by water currents. In this context, it is relevant here to point out the great importance of wave action, to which intertidal invertebrates, including many sea anemone species, are exposed.

How strong do the water currents have to be? The

293

literature varies on this. Average flow rates across a tropical reef flat may range from 2 to 4 cm (0.8-1.6 inches) per second. Yet, in wind exposed areas, such as the windward side of a reef, there may be an average current flow of 10 cm (4 inches) per second or 360 m (1188 feet) per hour. This, of course, is directional flow, which changes as tidal flows change.

Because of the rather limited dimensions of a home aquarium as compared to the vastness of the ocean, marine aquarists have to work with total water turnover per given time unit. The recommended range is usually five to ten changes or complete turnovers per hour. So, for instance, a tank with a volume of 300 liters (78 gallons) should be serviced by a circulating pump with a capacity of 1,500 to 3,000 liters (400-800 gallons) per hour. The discharge of this pump preferably should be directed into the tank in such a manner as to produce a circular current (which then approximates directional flow in the ocean).

Very large tanks in excess of 1 cubic meter = 1,000 liters (260 gallons) should ideally be equipped with two circulating pumps with discharge lines into opposite directions from each other. These pumps can operate alternately for about six hours each and so simulate very roughly a tidal flow situation.

In respect to the direction of water current, it should be noted here that wave action also is of critical importance in a miniature reef aquarium, particularly for intertidal invertebrates such as sea anemones. This can easily be simulated in an aquarium through the use of a dump bucket arrangement. The bucket pivots around a certain point above its center of gravity. It is then filled at a constant rate with water from the tank via a separate circulating pump. Once the pivoting level inside the bucket has been exceeded, the water will be dumped into the tank, creating breaker turbulence that rapidly subsides. The bucket then returns automatically to its resting position to be refilled automatically.

this sounds primitive, it is extremely effective in terms of animal well-being. Of course, these counter-weights will have to be on opposite sides of the tank and well out of view from the front of the tank. A small section on

Anemones such as this *Heteractis magnifica* obtain their food in nature by trapping tiny animals carried in the currents. Advanced hobbyists who attempt to duplicate these currents in the aquarium find that their animals often do better and reproduce more often. Photo: U. E. Friese.

If aquarists find such a system a bit cumbersome and messy (yes, it can splash a bit!), wave action on a smaller scale also can be accomplished by alternately lowering and raising a brick or cinder block into the tank. While

each side of the tank should be fenced off by a perforated plastic or glass panel from the main center section of the tank. The weights, suspended from a bar above the surface, are tilted alternately by an electric

Facing page: Anemones are not always the most colorful invertebrates in the aquarium. Featherduster worms here provide the color to set off a group of flame gobies, *Nemateleotris decora.* Photo: Dr. D. Terver.

motor drive. As one counterweight is submerged the other emerges from the water, creating very effective water surges from one side of the tank to the other.

Fishes as well as invertebrates usually react quite favorably to such wave action. Some of the invertebrates clearly depend on it; experiments have shown that certain intertidal anemones (e.g., *Oulactis muscosa,* which lives just below the low tide mark along the rocky shores off Sydney, Australia) require wave action for their survival. Specimens taken into an aquarium gradually die off without wave action, but with it this species displays long-term survival and even substantial growth in captivity (Taronga Zoo Aquarium).

The technology to create currents and wave action in a miniature reef aquarium is relatively simple. Circulating pumps are available in all sorts of sizes and capacities. Whatever the design, special care has to be taken on the intake sides of the pump, where there can be considerable suction. In order to prevent animals from being sucked into the pump (causing it to become blocked or sustain impeller damage), a suitable fine-meshed strainer must be placed over the inlet. It may also be advisable—for large tanks to achieve optimum water clarity—to connect a power filter vessel (filled only with gravel or filter wool) in front of the circulating pumps. This will act not only as an effective particle remover, but in time it also will provide additional biological filtration.

IMPORTANT DON'TS

A miniature reef aquarium—whether it is intended only for sea anemones or for sea anemones together with other invertebrates or with clownfishes—is basically a dynamically balanced but still natural system; at

Facing page: A view into the disc of the tube anemone *Cerianthus* shows the contrasting sizes of the tentacles typical of the group. Such specialization usually correlates with interesting oddities in prey selection. Photo: K. Lamberton.

least that is what marine aquarists should be aiming at. Nothing can upset this balance quicker than wrong methodology (doing the wrong things) or inappropriate technology (using wrong equipment).

There is some equipment, often widely used in conventional marine aquaria, that is inappropriate for a miniature reef tank. First, no protein skimmer. While these devices are very effective in aquarium water management, they have no place in the miniature reef aquarium. Protein skimmers, when working correctly, remove certain organic molecules that adhere to the bubbles formed by sea water when strongly aerated. Unfortunately, protein skimmers are not selective when it comes to deciding which organic molecules to remove and which to leave behind. While some of these can not be easily removed by other means, there are also those that are easily removed that normally would enter the nitrogenous waste product breakdown cycle. This, of course, is intended to take place in our trickle filter, so a protein skimmer could deprive the trickle filter of some of the nutrients required for the *Nitrobacter* and *Nitrosomonas* bacteria resident there. In other words, the trickle filter would function less effectively and not operate as efficiently as it should.

It must be noted here that some authors disagree with this viewpoint. They feel that a properly controlled and monitored protein skimmer is very much a part of a miniature reef aquarium. To a certain point, I agree with their views, but the test equipment needed to accurately monitor the effectiveness of a protein skimmer and at the same time maintain proper water quality often is beyond the financial means and experience of aquarists. So, I remain convinced that doing

without the protein skimmer is to the advantage of a miniature reef aquarium.

Second, ozonizers as well as ultraviolet lamps are principally intended to destroy undesirable bacteria. Ozone is a strong oxidizing agent that not only kills bacteria, viruses, and many protozoans, but also oxidizes many pollutants. Ultraviolet affects primarily bacteria and viruses, but does not seem to affect most bacterial spores. It also causes chemical changes in some pollutants. Consequently, in a miniature reef tank both of these devices could impede the proper functioning of the trickle filter. Moreover, if a miniature reef tank is well established and works the way it should, there really is no need for either of these devices. It must also be noted here that both ozone and ultraviolet radiation are highly dangerous to humans, so special precautions have to be taken to avoid human exposure.

Activated carbon is another substance that really has no applicability to a properly maintained and functioning miniature reef aquarium. Because of its ability to selectively remove certain chemical compounds from the water through the process of adsorption, it has been very popular as a filter medium in aquarium water management. But one of its disadvantages is the relatively short life of its adsorption capability. If it is not replaced in time, some of the substances already adsorbed on the charcoal will go back into solution or will be decomposed by bacteria. Consequently, some of the undesirable substances then are returned to the aquarium water. Because of this inherent unreliability, activated carbon is best left out of the miniature reef aquarium system.

Finally, a word about air stones (aeration). The fundamental principle of the miniature reef aquarium is an approximation of a dynamic equilibrium, where all major components are in a state of interactive balance. The miniature reef aquarium is able to generate and maintain its own oxygen saturation through the process of photosynthesis of higher algae. Providing additional aeration to the tank would mitigate this. While an air stone probably would do no harm, it also would not make a significant contribution to a well-functioning miniature reef aquarium and is best omitted. If dissolved oxygen measurements or the behavior of the animals indicates an oxygen shortage in the water, every effort should be made to find out why: tank is over populated, too much food given, inadequate trickle filtration, not enough light for proper photosynthetic activity, etc.

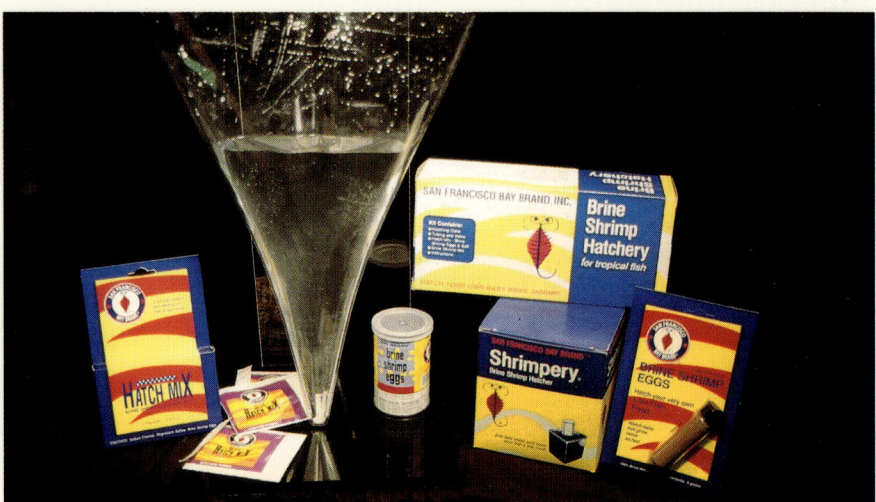

Brine shrimp eggs and all of the equipment and supplies needed for their successful hatching are widely available at pet shops. Photo courtesy of San Francisco Bay Brand.

Anemone Care

FEEDING SEA ANEMONES

In the wild, sea anemones feed opportunistically and more or less continually. In other words, they feed whenever there is food available. As part of sound aquarium management practices they will have to be adapted to feed at regular intervals. Apart from that, feeding sea anemones in an aquarium rarely presents a problem.

Most species feed willingly, provided the right kind and size of food are provided. Some species may even get by on very little food. Those species with substantial intracellular zooxanthellae aggregations seem to get by without food, provided the algae are able to maintain their normal photosynthetic activity with high light levels, which then also produces nutrients for the host anemone. For instance, the colonial mushroom anemones of the genera *Rhodactis* and *Actinodiscus,* as well as the small jewel anemones, *Corynactis* species, rarely need supplementary feeding as long as proper lighting (including actinic 03) is provided.

It must be emphasized at this point that light is absolutely essential for optimal growth of these symbiotic sea anemones. Consequently, the wave length of the lighting provided, together with the duration and intensity, can be used to provide optimal conditions for sea anemones. Generally speaking, sea anemones fall into two categories: those that are typical predators, and those that feed on free-floating organic particles and zooplankton.

Sea anemones with long, stout tentacles and with strong nematocyst action ("sticky") are predators. These include many popular aquarium types, such as *Tealia, Anemonia, Anthopleura,* and the giant tropical stichodactylid

A tube anemone surrounded by brittle stars. Photo by Courtney Platt.

anemones *Stichodactyla* and *Heteractis*. In their natural habitat these anemones feed entirely on small (and not-so-small) fishes and various invertebrates that come into direct contact with the powerful tentacles. In captivity they readily accept pieces of raw fish (fresh or frozen), crab, shrimp, and mussel meat, and similar items. Substitutes such as small pieces of raw beefheart and liver are also often taken. For efficient feeding the food should be placed (by hand or feeding stick or forceps) directly on the tentacles, which then will quickly manipulate the food into the mouth opening.

The size of food items for predatory sea anemones seems to be of very little importance. Observations in public aquaria, where sea anemones often are part of comprehensive community tank exhibits, have shown that they may start to ingest a whole fish (usually a sick or dying

Facing page: If you keep cnidarians you probably will need a food prepared specifically for invertebrates. Many corals and anemones are particle feeders that take only the finest floating foods from the current. Photo courtesy Coralife, Energy Savers.

specimen) several times the size of the sea anemone. The marine aquarist will find it more practical to offer food in small pieces, about 5 to 8 mm (0.2-0.3 inch) cubes for small specimens and 10 to 15 mm (0.4-0.6 inch) cubes for larger ones.

Once a piece of food has been placed on the tentacles, a healthy sea anemone will immediately invert the tentacular crown over the oral opening. In some species only part of the tentacles will invert, so additional pieces of food can be placed on those tentacles still open. Usually one to two pieces of food twice or three times a week are more than adequate. It is hard to put too much stress on this point, as over feeding definitely is one of the most important reasons for the failure of a miniature reef aquarium; excess food can quickly be converted into death-dealing pollution. Hobbyists *must* resist the temptation to feed more than is needed.

The ingested food particles not digested usually are regurgitated through the mouth within about 24 hours. This material must be removed immediately since it would foul the water very quickly, and in a miniature reef aquarium this could lead to a total system break-down. Also remove pieces of food rejected by the anemone and those that have fallen off the tentacles. Incidentally, food should be offered only when the anemones are fully expanded with all tentacles extended.

Particle feeding sea anemones are primarily those with thin, fine tentacles, such as *Metridium, Halcampa, Sagartia,* and others. In their natural habitat these anemones invariably feed on a variety of zooplankton, free-floating organisms from protozoans to various invertebrate larval forms to fish eggs and similar small items. Marine aquarists can substitute for planktonic food quite readily with newly hatched brine shrimp larvae *(Artemia). Artemia* larvae are readily accepted by most plankton-feeding sea anemones. They are an ideal food, since excess larvae will stay alive in the tank and do not foul the water.

In order to provide a variable diet, cultured marine protozoans (such as *Euplotes,* which is available through biological supply houses) make good sea anemone food. Some of the particle feeding sea anemones *(Metridium)* often accept very finely chopped solid food, such as macerated shrimp and fish flesh. Reasonably good food substitutes are some of the tiny freshwater crustaceans, such as water fleas and copepods. These, of course, have only a short life span in sea water, so they must be fed sparingly to avoid problems if they die and then decompose.

Many sea anemones can live in an aquarium for weeks without direct supplementary feeding. The situation is even less critical for those species that can rely on the photosynthetic activity of intracellular zooxanthellae for some of their nutrients. For these anemones the only essential factor is adequate illumination.

A final word of caution against overfeeding! In an aquarium situation this is dangerous at the best of times, particularly in a marine tank. It is even more critical in a miniature reef aquarium. Sea anemones—especially the predatory types—will take food far more often and in greater quantities than they really need. If sea anemones are fed more often than required for their normal metabolic activities, it will lead to an abnormal build-up of metabolic (nitrogenous) waste materials in the water. This, together with any uneaten food that is not promptly removed, can place an excessive burden on the entire system. The trickle filter may no longer be able to cope with the rapidly accumulating ammonia and nitrite (requiring more and more dissolved oxygen for the filter bacteria to do their work), and the subsequent increase in fertilizer will lead to a dramatic increase in the growth and spread of lower forms of algae in the tank. It does not take much to seriously unbalance a miniature reef aquarium, particularly one that accommodates primarily sea anemones. It is imperative that sea anemones are always fed in moderation.

ENEMIES OF SEA ANEMONES

Sea anemones have many enemies, but none greater than man! First, there are the crippling effects of man-made pollution on the inshore marine environment. Recent oil spills in Alaska and the English Channel, for instance, have killed untold millions of intertidal marine invertebrates; a great many of these were sea anemones. Large aggregations of sea anemones could once be seen at low tide above and below the water line on harbor wharves and pilings of many port facilities in some of the great sea ports of the world. Now there are mostly murky, smelly harbors, more or less devoid of any marine life.

Even coral reefs in the middle of the South Pacific and Indian Ocean have had their invertebrate and vertebrate faunas decimated by the effects of heavy metal and pesticide pollution carried around the oceans by global currents.

Increasing human populations that collectively have more and more leisure time on their hands also produce constantly increasing crowds at local and exotic sea shores. There they cause untold harm because of careless or unintentional destruction of the habitat of intertidal animals. Sea anemones that are prominent in their appearance even when exposed at low tide often are mutilated by idle beachgoers. Rocks in the intertidal zone are turned over at low tide in search of interesting animals, without being returned to their previous position.

Consequently, sunlight, dehydration, and general exposure will kill off the multitude of invertebrates on the underside of the rock. In some locations many of these are small sea anemones.

The destruction is particularly serious on coral reefs. Many of the giant tropical sea anemones live just below the surface, and there they are often mercilessly trodden on by careless reef walkers, who beat a destructive path through the shallow water coral wonderland.

Man is also a predator of sea anemones in even more direct ways. First, there is the collecting of sea anemones by an ever-increasing number of marine invertebrate collectors. Many of the animals taken, including large numbers of sea anemones, are removed by professional collectors to supply the marine aquarium trade. Then there is the scientific and teaching community that requires animals for

309

research and education. Again, often it involves sea anemones.

Sea anemones also are eaten by man. Fishermen along the Mediterranean coast have for hundreds of years collected sea anemones as gourmet delicacies. Even relatively recent literature suggests, for instance: *Actinia equina:* "Edible—dipped into batter, deep-fried in oil and served with tartar sauce." There are similar recipes for *Anemonia sulcata* and *Calliactis parasitica!*

Sea anemones have a number of natural enemies that feed on them or kill them in the constant battle for living space. Small bottom-dwelling sea anemones are eaten by many benthic fishes (flatfishes, cod, some eels, and

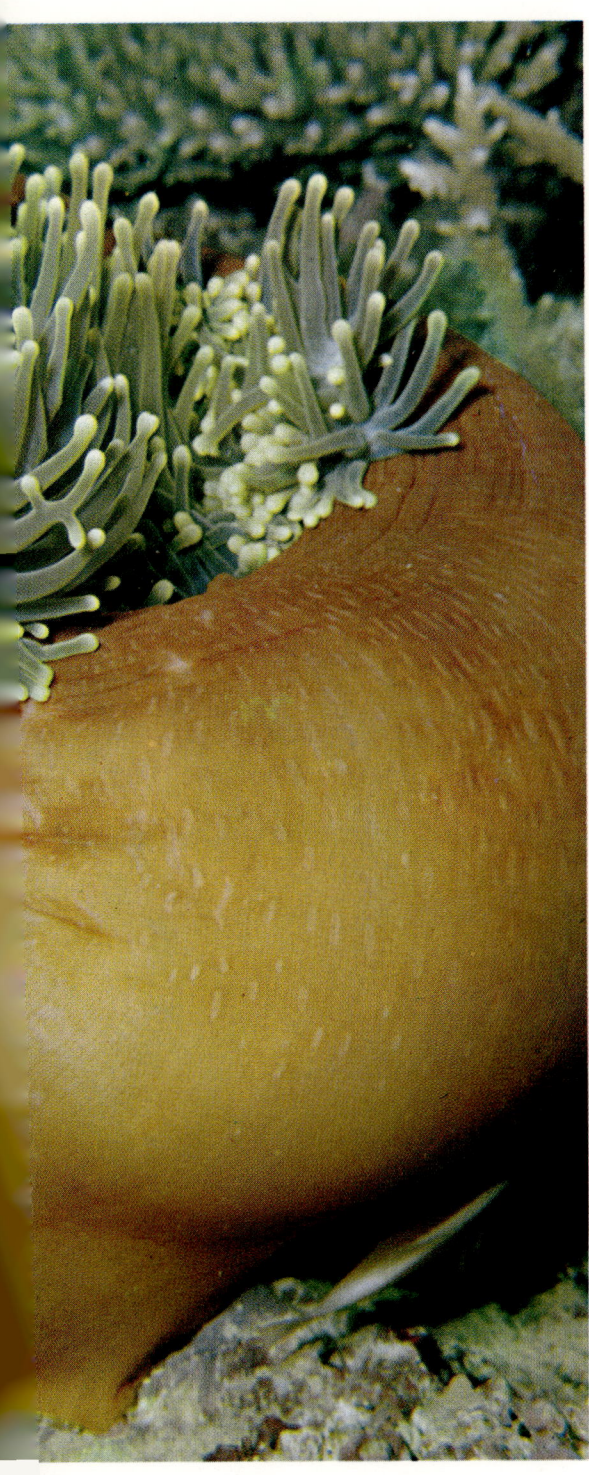

many others). Certain starfish and molluscs are also known predators of sea anemones. There often also is fierce competition among sea anemones for suitable sites of attachment. The colorful tropical mushroom anemones of the genera *Actinodiscus* and *Rhodactis* have a tendency to inhibit the spread of other species. The small, semi-transparent *Aiptasia (A. diaphana, A. pallida, A. luciae)*—often introduced accidentally into a tank together with other animals—often exhibit almost explosive reproduction and growth and quickly out-compete most other sea anemones.

A few sea

Good enough to eat! In some regions, fishermen actively collect sea anemones and prepare them for market. Some species have a high reproductive rate, and it is not unthinkable to visualize anemone farms in the future. Photo of *Heteractis magnifica*: A. Power.

Facing page: Although some crabs, such as this *Petrolisthes ohshimai*, occur as commensals in anemones, most crabs are dangerous, being able to take food from anemones and damage their delicate tentacles and disc with the crabs' heavy bodies and pointed legs. Photo: A. Norman.

anemone species must never make physical contact with each other, because this could be fatal for one or both of them. The column of *Actinia equina* is equipped with pronounced sac-like protuberances that are heavily armed with nematocysts. If another *A. equina* specimen (or some other species) approaches the one already in place, the resident will inflate these sacs with water and actually lean toward the newcomer. The skin of the sacs will adhere to the column of the newly arrived sea anemone and the nematocysts will cause considerable tissue destruction. This often leads to the demise of the attacked sea anemone unless it retreats in time.

On the other hand, sea anemone species with similar marginal sacs *(Anemonia sulcata* and *Bunodactis)* appear to be much more tolerant, certainly toward their own kind, and will permit some physical contact with each other.

In a marine aquarium, sea anemones as well as other inhabitants of the same tank have to be carefully selected for mutual compatibility. If this is not done, problems can arise quickly. Triggerfishes and moray eels, for instance, are known to molest newly introduced sea anemones. Small specimens may be swallowed whole (often mistaken for food) and then spat out again severely injured, while larger specimens are often badly mauled by these aggressive fishes.

Another difficulty with sea anemones in a normal community marine tank arises from the fact that many fishes will steal pieces of food given to sea anemones. This problem is particularly acute with large, aggressively feeding fishes that usually are (because of their size) largely immune to the stings of nematocysts. Not only will large fishes take food away from sea anemones, but so will large crabs. Their hard

carapaces are virtually impregnable to even the most powerful nematocysts of sea anemones. Moreover, the large and clumsy walking legs and the strong pincers of crabs can severely damage a sea anemone.

In captivity, sea anemones ideally are kept in a miniature reef aquarium. There they should be an integral part of an invertebrate community, with emphasis on corals, polychaete worms, small crustaceans, molluscs, and possibly a few fishes. A word of caution here: remember the predatory habits of some sea anemones! It usually is safest to select co-inhabitants from among those invertebrates that are known to live compatibly together, especially those that are known symbionts of sea anemones.

A final word about the ultimate enemy of sea anemones in the home aquarium: faulty technology! Wrong procedures and inadequate or faulty equipment are the most common causes of the demise of sea anemones in captivity. Poor quality sea water (inadequate biological filtration); chemical pollution (aerosol home insecticides are usually highly water soluble and very toxic to all marine organisms); ozone (highly toxic); and inadequate heaters or refrigeration units all take their toll on anemones.

Temperature equipment failure is the most frequent cause for accidental sea anemone deaths. Sea anemones are either "cooked" or "frozen" to death. In this context it may be very pertinent to point out again that all sea anemone species must be maintained at temperature ranges compatible with those occurring in their natural habitat.

How To Get a Sea Anemone

BUYING SEA ANEMONES

Sea anemones are cosmopolitan creatures that occur in all oceans, from the icy waters of the far north and far south, to the temperate latitudes of the central Atlantic and Mediterranean Sea, to the tropics of the central Indian and Pacific Oceans and all adjacent seas. Sea anemones, especially those from the Mediterranean Sea, southern Florida and the Caribbean, and the tropical Indian and Pacific Oceans, have been available commercially for many years. Most major aquariums shops have good selections of marine invertebrates, including sea anemones. So, do we buy our sea anemones or do we put on rubber boots or SCUBA gear and collect our own?

Clearly, the easy way is to purchase sea anemones. Not everybody has quick and easy access to the seashore, and few of us live in the tropics where some of the more exotic sea anemones are found. If the choice has been made to purchase sea anemones for our newly established miniature reef aquarium, what do we look for when we see sea anemones on display at the local aquarium store?

With few exceptions, it is inherently difficult to tell whether a sea anemone is healthy or not. A specimen that looks and behaves perfectly normal one day may decline in condition rapidly overnight and be dead within a couple of days. Nevertheless, there are certain signs that a cautious aquarist should look for when he or she acquires sea anemones.

Like all marine invertebrates, sea

anemones MUST have clean water to live in. How does the dealer keep his sea anemones? Are his tanks clean and free of accumulated organic debris and left-over food? Is the water perfectly clear? If the answer is *no* to any of these questions, the aquarist should be very cautious. Frankly, I would always prefer to go to another dealer! Sea anemones must have perfectly clean surroundings, and specimens kept under less than desirable conditions are bound to be affected by this, although there may not yet be any definite external signs that a specimen is unwell.

Often the question "When did these sea anemones come in and where from?" provides a general idea about how long the animals have already been in captivity and how much time they must have spent in transit. Both of these conditions will have some effect on sea anemones. Moreover, since there are inevitable in-transit and/or transport losses that do not materialize until AFTER (often some time after) the animals have arrived, it is important to know whether the animals now on display could still be under recent shipment stress or whether they are already survivors.

Provided the sea anemones on display at the dealer have had an opportunity to become acclimated and are settled in (and are obviously being kept under acceptable conditions), the specimen(s) tentatively selected should be closely examined in the tank. Is the animal open? Are all tentacles fully expanded, their tips not shriveled? Is the column fully inflated, giving a strong, stout and healthy appearance? If this is so then there is a fairly good chance that this sea anemone MAY BE healthy. But it also should be remembered that even perfectly healthy specimens at times are limp and fully contracted. Nevertheless, a certain caution must be exercised with specimens that are not fully open.

Index

(Page numbers in bold face refer to illustrations.)

Actinia, 73, 78, **106**
Actinia ehrenbergi, 234
Actinia equina, 55, 63, **64, 66,** 68, 70, 71, **129, 130,** 131, 132, 310, 312
Actinia equina japonica, 129
Actinia equina pontica, 129
Actinia helianthus, 234
Actinia tenebrosa, **65, 131,** 132, **133**
Actinodiscus, 76, 115, **115,** 116, **116-117,** 117, **120,** 248, 304, 311
Actinodiscus numiformis, 116
Actinostola abyssorum, 178
Actinostola callosa, 178
Adamsia, 182
Adamsia palliata, 211
Aeolidia papillosa, 62
Aiptasia, 68, **170, 195, 197,** 290
Aiptasia couchii, 196
Aiptasia diaphana, 196, 311
Aiptasia luciae, 311
Aiptasia mutabilis, 196
Aiptasia pallida, 198, 311
Aiptasia tagetes, **198,** 198
Amphiprion, 76, 148, 216, 224, 226, 235, 236, 239, 242,
Amphiprion akallopisos, **221**
Amphiprion akindynos, **51, 233**
Amphiprion allardi, **212-213, 220**
Amphiprion bicinctus, **172**
Amphiprion chrysopterus, **9, 59, 179, 218, 232,** 242
Amphiprion clarkii, **54, 57, 175, 176-177, 227,** 234, **235,** 242, **240-241,** 244, 248
Amphiprion leucokranus, **48**
Amphiprion melanopus, **80-81, 228-229**
Amphiprion nigripes, **240-241, 286-287**
Amphiprion ocellaris, **11, 31, 36**
Amphiprion percula, 222

Amphiprion perideraion, **45, 112-113, 249**
Amphiprion tricinctus, 242, **252-253**
Amphiprion xanthurus, 131
Anemonia, 56, 73, 78, 82, 83, 304
Anemonia sulcata, 64, **70, 71, 83, 142,** 142-143, 146, 150, 214, 310, 312
Anthelia, 290
Antheopsis, 217, 218
Anthopleura, 304
Anthopleura (=Aulactinia) crassa, 154
Anthopleura artemisia, 153
Anthopleura ballii, 154
Anthopleura elegantissima, 55, 71, 150-154
Anthopleura japonica, **150**
Anthopleura midori, **151**
Anthopleura thallia, 154
Anthopleura xanthogrammica, 71, **72, 88,** 88, 149-150, **153, 154**
Anthothoe, **185, 186**
Anthothoe albocincta, 183-184, **184**
Antiparactis, **47**
Artemia, 146, 194, 307
Atlantic Tube Anemone, 94
Australian Jewel Anemone, 110
Australian Speckled Anemone, 154
Australian Swimming Anemone, 167
Banded Sea Anemone, 165
Bartholomea annulata, 196, **200, 201, 203**
Beaded Anemone, 165
Bolocera tuediae, **125**
Boloceroides, 123
Bottle-green Anemone, 158
Bunodactis, 31, 64, 73
Bunodeopsis, 123
Bunodeopsis prehensa, 124
Bunodeopsis strumosa, 124

317

Buying anemones, 315-316
California Jewel Anemone, 110
Calliactis parasitica, **183,** 210, 310
Calliactis, 67, 73, 180-182, **181,** 211, 212, **219, 225,** 226
Cassiopea, **24**
Cassiopea frondosa, **26-27**
Caulerpa, **112-113, 281**
Cavernularia obesa, **205,** 206
Centropyge acanthops, **112-113**
Cerianthemorpha brasiliensis, 96
Ceriantheopsis americanus, 94-97, **97**
Cerianthus, **58, 61, 91,** 92, 126, 226, **272, 299**
Cerianthus borealis, 94, **95,** 96
Cerianthus lloydii, 94, **96**
Cerianthus membranaceus, **93,** 93-94, 98
Chalky Sponge Anemone, 105
Chironex fleckeri, **43**
Circulation, in aquarium, 290-296
Cnidopus verater, 158-160, **160**
Condylactis, **10, 32, 36,** 68
Condylactis aurantiaca, **143,** 143-146, 214
Condylactis gigantea, 146-149, **147,** 216
Condylactis passiflora, **144, 146, 148, 149**
Corynactis, **37,** 304
Corynactis australis, 108, 110
Corynactis californica, **109,** 110-111, **111**
Corynactis viridis, 110
Crimson Sea Anemone, 129
Cryptodendrum adhaesivum, 234
Crystal Anemone, 123
Cyanea capillata, **34**
Dardanus sp., **219**
Dardanus arrosor, **183,** 211
Dascyllus, 76
Dascyllus albisella, 248
Dascyllus trimaculatus, 235, **236, 239,** 239, **240-241,** 244, 246, 248, **250-251**
Delicate Anemone, 187
Dermasterias imbricata, 60, **62**
Discosoma, 116, 117

Discosoma gigantea, 244
Dofleinia, 165-166, **166, 167**
Eastern Pacific Tube Anemone, 98
Ecklonia radiata, 167
Edwardsia, 56, 69, 126
Edwardsia andresi, 126
Edwardsia californica, 127
Edwardsia leidyi, 127
Edwardsia lineata, 127
Edwardsia longicornis, 126
Edwardsia sipunculoides, 127
Encrusting anemones, 99
Enemies, of anemones, 308-314
Entacmaea, 216
Entacmaea quadricolor, **40, 53, 57, 80-81, 84, 172, 215, 228-229, 233, 234**
Epizoanthus arenaceus, 100
Euplotes, 307
European Jewel Anemone, 110
European Wax Anemone, 142
Filter, trickle, 267, **267, 268**
Foods and feeding, 73-78
Giant Green Anemone, 149
Glass anemones, 194
Gobiodon rivulatus, **145**
Gobius bucchichi, 143
Golden Anemone, 143
Golden Mat Anemone, 101
Gonactinia prolifera, 68, 123
Goniopora, **236,** 290
Green Mat Anemone, 101
Halcampa, 56, 74, 75, 78, 82, 126, 306
Halcampa duodecimcirrata, 127
Halcampoides purpurea, 126
Haliclystus, **25**
Heliofungia sp., **246-247**
Hermit anemones, 180
Heteractis, 114, 216, 236, **250-251,** 305
Heteractis aurora, 244, **252-253**
Heteractis crispa, **8, 9, 45, 48, 218, 232, 237, 242, 243,** 244
Heteractis magnifica, **7, 11, 16-17, 31, 51, 59, 227, 240-241,** 244, **245, 246, 249, 289, 295, 310-311**
Holacanthus tricolor, 102

Internal anatomy, 46-54
Iotrochata birotula, 102
Jewel anemones, 108
Knobby Jewel Anemone, 111
Lebrunia coralligens, 162, **163**
Lebrunia danae, 160-165, **162, 163**
Lighting, in aquarium, 278-283, **280, 282, 283, 284, 301**
Lima scabra, 216
Little Striped Anemone, 183
Lybia, 124
Lybia tessellata, **211, 217**
Macrodactyla doreensis, 235
Macropodia rostrata, 143
Mantle anemones, 182
Mediterranean Mat Anemone, 105
Mediterranean Tube Anemone, 93
Metridium, 56, 67, 71, 78, 82, 83, 306, 307
Metridium marginatum, **79, 190**
Metridium senile, 55, 188-194, **191, 193**
Minature reef aquarium, history, 258-261
Minature reef aquarium, setting up, 263-264
Minyas, 60, 170
Movement and locomotion, 67-69
Murex, 180
Mushroom anemones, 114
Mushroom Mat Anemone, 102
Nautactis, 170
Nemateleotris decora, **296**
Neopetrolisthes, 214
Nitrobacter, 269, 298
Nitrosomonas, 269, 298
North American Tube Anemone, 94
Northern European Tube Anemone, 94
Oulactis, 56
Oulactis muscosa, 154-158, **155, 156, 157, 159,** 160, 296
Pachycerianthus estuari, 98
Pachycerianthus johnsoni, 98
Pachycerianthus torreyi, **98**, 98
Pagurus, 182
Pagurus prideauxi, 211
Pagurus bernhardus, 211

Palythoa, 290
Palythoa caribbea, 101-102
Palythoa grandis, 102
Parazoanthus, **52**, 56, 289
Parazoanthus axinellae, 100, **104, 105**
Parazoanthus parasiticus, 105
Parazoanthus swiftii, 102
Peachia, 69, 73, 82
Peachia hastata, 126, **127**
Pelagia noctiluca, **23**
Pennatula aculeata, 204
Pennatula phosphorea, 204
Periclimenes, **215,** 234, 236, 237, 244
Periclimenes brevicarpalis, 214
Periclimenes holthuisi, 214
Periclimenes pedersoni, 149, 162
Periclimenes scriptus, 143
Petrolisthes maculatus, 214, **259**
Petrolisthes ohshimai, 214, 244, **313**
Phlyctenactis tuberculosa, 167-170, **168, 169**
Phlyctenanthus australis, 168, **171**
Phymanthus, **69**
Phymanthus crucifer, **128, 164, 165,** 165
Physalia, **20**
Pink-tipped Anemone, 146
Polydectus cupulifer, **210**
Population size, in aquarium, 286-290, **291**
Protanthea, 123
Protection, by anemones, 60-67
Pseudactina flagellifera, **152**
Pseudocorynactis, **46**
Pteroides spinosum, 204
Ptilosarcus gurneyi, 206, **208**
Radianthus gelam, 234
Radianthus malu, 244, 248
Radianthus paumotensis, 244
Radianthus ritteri, 244
Radianthus simplex, 244
Reproduction and development, 78-83
Rhodactis, 76, 116, 117, **119,** 248, 304, 311
Rhodactis howesi, **77**
Rhodactis sanctithomae, 116

Ricordea florida, **105,** 111-114
Rough Sea Anemone, 150
Sagartia, 74, 306
Sagartia davisi, 67
Sagartia rhododactylus, 187-188
Sagartia troglodytes, 78
Sagartian anemones, 187
Sensory responses, 69-72
Stichodactyla, **35,** 74, 114, 216, **238,** 305
Stichodactyla gigantea, **179,** 237, 238, 239
Stichodactyla haddoni, **212-213,** 242, 244, **259**
Stichodactyla helianthus, **12, 73, 74, 230,** 236, 237
Stichodactyla mertensii, **75,** 239, 242
Stichodactyla tapetum, 236
Stinging Branched Anemone, 160
Stoichactis, 68, 73
Stoichactis kenti, 242
Stomphia, 62, **62, 63,** 67, 68, 71
Stomphia coccinea, 60, **68,** 170, 174-178
Stylatula elongata, **206,** 206
Swimming Anemone, 174
Tealia, 56, 67, 73, 134, 136, 137, 138, 140, 141, 304
Tealia columbiana, **41, 134, 135,** 140
Tealia coriacea, **87,** 88, **89,** 132-142, **137,** 139, 140
Tealia crassicornis, **136**
Tealia felina, 71, 78, **139,** 139, 140
Tealia lofotensis, **82, 86, 138,** 140
Tealia piscivora, **141**
Telmatactis decora, **210**
Temperature range, in aquarium, 283-285
Thor, 234, 237
Thor amboinensis, 149
Triactis, **217**
Triactis producta, 124, **211**
Tube anemones, 91
Tubular or Beaded Sea Anemone, 132
Veretillum cynorium, 206
Waratah Sea Anemone, 132
Water currents, in aquarium, 290-296
White-plumed or Piling Anemone, 188
Xenia, 290
Yellow Sponge Anemone, 102
Zoanthus sociatus, 101